CURRENT CONCEPTS OF A NEW ANIMAL MODEL: THE NON MOUSE

CURRENT CONCEPTS OF A NEW ANIMAL MODEL: THE NON MOUSE

Edited by

N. Sakamoto, N. Hotta

Third Department of Internal Medicine, Nagoya University School of Medicine, Nagoya, Japan

and

K. Uchida

Aburahi Laboratories, Shionogi & Co., Ltd., Shiga, Japan

1992
ELSEVIER
AMSTERDAM · LONDON · NEW YORK · TOKYO

ELSEVIER SCIENCE PUBLISHERS B.V.
Sara Burgerhartstraat 25
P.O. Box 211, 1000 AE Amsterdam, The Netherlands

Library of Congress Cataloging-in-Publication Data

Current concepts of a new animal model : the NON mouse / edited by N. Sakamoto, N. Hotta, and K. Uchida.
p.- cm.
Includes bibliographical references and index.
ISBN 0-444-89554-X (alk. paper)
1. Non-insulin-dependent diabetes--Animal models. 2. Kidney--Diseases--Animal models. 3. Mice as laboratory animals.
I. Sakamoto, N. (Nobuo), 1931- . II. Hotta, N. III. Uchida, K. (Kiyohisa)
[DNLM: 1. Diabetes Mellitus, Insulin-Dependent. 2. Kidney Diseases. 3. Mice, Inbred Strains. 4. Models, Biological. QY 60.R6 C976]
RC660.C84 1992
616.4'62027--dc20
DNLM/DLC
for Library of Congress 92-14878
CIP

ISBN: 0 444 89554 X

This book is printed on acid-free paper.

Printed in The Netherlands.

List of Contributors

T. Chikai
Shionogi Research Laboratories, Shionogi & Co., Sagisu, Fukushima-ku, Osaka 553, Japan.

S. Fukushima
First Department of Pathology, Osaka City University Medical School, 1-4-54 Asahimachi, Abeno-ku, Osaka 545, Japan.

T. Hanafusa
Second Department of Internal Medicine, Osaka University Medical School, 1 Chome, Fukushima-ku, Osaka 553, Japan.

M. Harada
Shionogi Research Laboratories, Shionogi & Co., Ltd., Sagisu, Fukushima-ku, Osaka 553, Japan.

G. Hasegawa
First Department of Internal Medicine, Kyoto Prefectural University of Medicine, Kajii-cho, Hirokooji agaru, Kawaramachi, Kamigyo-ku, Kyoto 602, Japan.

M. Hata
First Department of Internal Medicine, Kyoto Prefectural University of Medicine, Kajii-cho, Hirokooji agaru, Kawaramachi, Kamigyo-ku, Kyoto 602, Japan.

Y. Hayashi
Aburahi Laboratories, Shionogi & Co., Ltd., 1405, Gotanda, Koka-cho, Koka-gun, Shiga 520-34, Japan.

K. Higashino
Third Department of Internal Medicine, Hyogo College of Medicine, Mukogawa-cho, Nishinomiya 663, Japan.

M. Hosono
Aburahi Laboratories, Shionogi & Co., Ltd., 1405, Gotanda, Koka-cho, Koka-gun, Shiga 520-34, Japan.

N. Hotta
Third Department of Internal Medicine, Nagoya University School of Medicine, 65 Tsurumai, Showa-ku, Nagoya 466, Japan.

H. Igimi
Shionogi Research Laboratories, Shionogi & Co., Ltd., Sagisu, Fukushima-ku, Osaka 553, Japan.

H. Imura
Second Division, Department of Internal Medicine, Kyoto University School of Medicine, 54 Shogoin Kawahara-cho, Sakyo-ku, Kyoto 606, Japan.

Y. Itoh
Third Department of Internal Medicine, Nagoya University School of Medicine, 65 Tsuruma, Showa-ku, Nagoya 466, Japan.

H. Ishida
Department of Metabolism and Clinical Nutrition, Kyoto University School of Medicine, 54 Shogoin Kawahara-cho, Sakyo-ku, Kyoto 606, Japan.

T. Kanatsuna
Department of Metabolic Disease, Kyoto City Hospital, Nakagyo-ku, Kyoto 604, Japan.

K. Kataoka
Division of Endocrinology and Internal Medicine, Keio University School of Medicine, 35, Shinanomachi, Shinjuku-ku, Tokyo 160, Japan.

S. Kato
Department of Metabolism and Clinical Nutrition, Kyoto University School of Medicine, 54 Shogoin Kawahara-cho, Sakyo-ku, Kyoto 606, Japan.

Y. Kato
First Division, Department of Internal Medicine, Shimane Medical University, 89-1, Enya-cho, Izumo-shi, Shimane 693, Japan.

T. Kayahara
Third Department of Internal Medicine, Hyogo College of Medicine, Mukogawa-cho Nishinomiya 663, Japan.

N. Koh
Third Department of Internal Medicine, Nagoya University School of Medicine, 65 Tsurumai, Showa-ku, Nagoya 466, Japan.

M. Kondo
First Department of Internal Medicine, Kyoto Prefectural University of Medicine, Kajii-cho, Hirokooji agaru, Kawaramachi, Kamigyo-ku, Kyoto 602, Japan.

N. Kono
Second Department of Internal Medicine, Osaka University Medical School, 1 Chome, Fukushima-ku, Osaka 553, Japan.

K. Kunimoto
Aburahi Laboratories, Shionogi & Co., Ltd., 1405, Gotanda, Koka-cho, Koka-gun, Shiga 520-34, Japan.

T. Kurose
Department of Metabolism and Clinical Nutrition, Kyoto University School of Medicine, 54 Shogoin Kawahara-cho, Sakyo-ku, Kyoto 606, Japan.

S. Makino
Aburahi Laboratories, Shionogi & Co., Ltd., 1405, Gotanda, Koka-cho, Koka-gun, Shiga 520-34, Japan.

T. Maruyama
Division of Internal Medicine, Social Insurance Saitama Chuo Hospital, Kitaurawa, Urawa-shi 366, Japan.

S. Matsui
Kanzakigawa Laboratory, Shionogi Research Laboratories, Shionogi & Co., Ltd., 3 Chome, Futaba-cho, Toyonaka 561, Japan.

J. Miyagawa
Second Department of Internal Medicine, Osaka University Medical School, 1 Chome, Fukushima-ku, Osaka 553, Japan.

D. Mizumoto
Third Department of Internal Medicine, Nagoya University School of Medicine, 65 Tsurumai, Showa-ku, Nagoya 466, Japan.

S. Mori
Aburahi Laboratories, Shionogi & Co., Ltd., 1405, Gotanda, Koka-cho, Koka-gun, Shiga 520-34, Japan.

T. Murai
Aburahi Laboratories, Shionogi & Co., Ltd., 1405, Gotanda, Koka-cho, Koka-gun, Shiga 520-34, Japan.

Y. Muraoka
Kanzakigawa Laboratory, Shionogi & Co., Ltd., 3 Chome, Futaba-cho, Toyonaka 561, Japan.

N. Nakamura
First Department of Internal Medicine, Kyoto Prefectural University of Medicine, Kajii-cho, Hirokooji agaru, Kawaramachi, Kamigyo-ku, Kyoto 602, Japan.

K. Nakano
First Department of Internal Medicine, Kyoto Prefectural University of Medicine, Kajii-cho, Hirokooji agaru, Kawaramachi, Kamigyo-ku, Kyoto 602, Japan.

M. Nishimura
Institute for Experimental Animals, Hamamatsu University School of Medicine, 3600, Handa-cho, Hamamatsu 431-31, Japan.

Y. Nomura
Shionogi Research Laboratories, Shionogi & Co., Ltd., Sagisu, Fukushima-ku, Osaka 553, Japan.

K. Ogata
Division of Pathology, Keio University School of Medicine, 35, Shinanoma-chi, Shinjuku-ku, Tokyo 160, Japan.

S. Ohgaku
First Department of Medicine, Toyama Medical and Pharmaceutical University, 2630, Sugitani, Toyama 930-01, Japan.

T. Ohhara
Aburahi Laboratories, Shionogi & Co., Ltd., 1405, Gotanda, Koka-cho, Koka-gun, Shiga 520-34, Japan.

Y. Okamoto
Second Division, Department of Internal Medicine, Kyoto University School of Medicine, 54 Shogoin Kawahara-cho, Sakyo-ku, Kyoto 606, Japan.

H. Sahata
Shimane Institute of Health Science, Shimane Medical University, 89-1, Enya-cho, Izumochi, Shimane 693, Japan.

N. Sakamoto
Third Department of Internal Medicine, Nagoya University School of Medicine, 65 Tsurumai, Showa-ku, Nagoya 466, Japan.

T. Saruta
Division of Endocrinology and Internal Medicine, Keio University School of Medicine, 35, Shinanomachi, Shinjuku-ku, Tokyo 160, Japan.

T. Sawa
First Department of Internal Medicine, Toyama Medical and Pharmaceutical University, 2630 Sugitani, Toyama, Japan.

Y. Seino
Department of Metabolism and Clinical Nutrition, Kyoto University School of Medicine, 54 Shogoin Kawahara-cho, Sakyo-ku, Kyoto 606, Japan.

S. Suzuki
Institute of Experimental Animals, Shimane Medical University, 89-1, Enya-cho, Izumo-shi, Shimane 693, Japan.

H. Takase
Shionogi Research Laboratories, Shionogi & Co., Ltd., Sagisu, Fukushima-ku, Osaka 553, Japan.

I. Takei
Division of Endocrinology and Internal Medicine, Keio University School of Medicine, 35, Shinanomachi, Shinjuku-ku, Tokyo 160, Japan.

N. Takeuchi
The Central Laboratories, Ehime University Hospital, Shigenobu-cho, Ehime 791-02, Japan.

Y. Takeuchi
Aburahi Laboratories, Shionogi & Co., Ltd., 1405, Gotanda, Koka-cho, Koka-gun, Shiga 520-34, Japan.

T. Taminato
Second Department of Medicine, Hamamatsu University School of Medicine, 3600, Handa-cho, Hamamatsu 431-31, Japan.

K. Tanigawa
First Division, Department of Internal Medicine, Shimane Medical University, 89-1, Enya-cho, Izumo-shi, Shimane 693, Japan.

T. Tominaga
Second Department of Medicine, Hamamatsu University School of Medicine, 3600, Handa-cho, Hamamatsu 431-31, Japan.

K. Tsuji
Second Division, Department of Internal Medicine, Kyoto University School of Medicine, 54 Shogoin Kawahara-cho, Sakyo-ku, Kyoto 606, Japan.

K. Tsukahara
Aburahi Laboratories, Shionogi & Co., Ltd., 1405, Gotanda, Koka-cho, Koka-gun, Shiga 520-34, Japan.

Y. Tsuura
Second Division, Department of Internal Medicine, Kyoto University School of Medicine, 54 Shogoin Kawahara-cho, Sakyo-ku, Kyoto 606, Japan.

K. Uchida
Aburahi Laboratories, Shionogi & Co., Ltd., 1405, Gotanda, Koka-cho, Koka-gun, Shiga 520-34, Japan.

H. Wainai
Division of Endocrinology and Internal Medicine, Keio University School of Medicine, 35, Shinanomachi, Shinjuku-ku, Tokyo 160, Japan.

H. Watanabe
Kanzakigawa Laboratory, Shionogi & Co., Ltd., 3 Chome, Futaba-cho, Toyonaka 561, Japan.

Y. Watanabe
Third Department of Internal Medicine, Nagoya University School of Medicine, 65 Tsurumai, Showa-ku, Nagoya 466, Japan.

K. Yamamoto
Second Department of Internal Medicine, Osaka University Medical School, 1 Chome, Fukushima-ku, Osaka 553, Japan.

H. Yamashita
Aburahi Laboratories, Shionogi & Co., Ltd., 1405, Gotanda, Koka-cho, Koka-gun, Shiga 520-34, Japan.

T. Yamazaki
Second Department of Medicine, Hamamatsu University School of Medicine, 3600, Handa-cho, Hamamatsu 431-31, Japan.

F. Yoshida
Third Department of Internal Medicine, Nagoya University School of Medicine, 65 Tsuruma, Showa-ku, Nagoya 466, Japan.

T. Yoshimi
Second Department of Medicine, Hamamatsu University School of Medicine, 3600, Handa-cho, Hamamatsu 431–31, Japan.

Preface

On one winter day in 1966, a female mouse showing a cataract was found in outbred ICR mice at Shionogi Aburahi Laboratories (Shiga, Japan). This mouse became the first mother of the inbred CTS strain. During the process of establishing the CTS strain, an attempt to select euglycemic and hyperglycemic lines was made at the sixth generation. A diabetic female mouse arose at the 20th generation of the euglycemic line, which was later established as the inbred NOD (non-obese diabetic) mice. Now it is well known that NOD mouse is one of the excellent animal models for the study of insulin dependent diabetes in human subjects. At that time, another strain was selected as a control for the NOD mice from the slightly hyperglycemic line and this strain was later developed as the NON (non-obese non-diabetic) mice.

Although NON mice were used as the control strain for NOD mice, NON mice were found to show a defect in glucose tolerance and abnormality of the renal glomerulus. Thus a meeting was held at Tokyo on May 28th in 1988 to exchange the data on NON mice, and subsequently the second (April 22, 1989) and the third (May 25, 1990) meetings were held at Tokyo. The contents presented and discussed in these three meetings suggest that the NON mouse may become a suitable animal model for either insulin independent diabetes or renal dysfunction.

This volume summarizes the reports on studies using NON mice in the first few years and also the contents will no doubt reflect the excellent fruition of three meetings held at Tokyo over three years.

We would like to express our deep gratitude to all those who helped in arranging these three meetings, leading to a success.

Nobuo Sakamoto
Nigishi Hotta
Kiyohisa Uchida

Contents

Section III – Characteristics of the nephropathy in the NON mouse

SECTION I

Breeding and genetics

CHAPTER 1

Breeding of the NON mouse and its genetic characteristics

SUSUMU MAKINO, HIROFUMI YAMASHITA, KIKUKO KUNIMOTO, KIYOSHI TSUKAHARA and KIYOHISA UCHIDA

Aburahi Laboratories, Shionogi & Co., Ltd., Koka, Shiga, Japan

Current concepts of a new animal model: The NON mouse
Edited by N. Sakamoto, N. Hotta and K. Uchida

Contents

Introduction

In 1980, we established the non-obese diabetic (NOD) mouse characterized by hyperglycemia, glycosuria, polyuria, rapid weight loss, absence of insulin secretion and destructive insulitis as an animal model for Type I diabetes mellitus [1, 2]. We also developed three related strains from the same origin as the NOD mouse and named them non-obese non-diabetic (NON), cataract Shionogi (CTS) and noncataract (NCT). However, we could not obtain a control strain for the NOD mouse from the NOD family, so that the NON mouse which did not show glycosuria and insulitis served as the control strain for the NOD mouse.

In the study of the NON mouse as a control for the NOD mouse, we found that the NON mouse exhibits interesting characteristics such as glucose intolerance and unique renal lesions. At present, the glucose intolerance characteristics of the NON mouse are expected to attract attention as a model for Type II diabetes mellitus [3, 4]. In addition, the renal lesions of the NON mouse are considered to be a useful model of human lipoprotein glomerulopathy [5].

The present paper describes the breeding and genetic characteristics of the NON mouse.

Breeding of the NON mouse

The genealogy of the NON mouse and its related strains are shown in Fig. 1. Briefly, their history is as follows. Dr. Ohotori found a mouse with cataracts and small eyes among outbred ICR mice purchased from Clea Japan, Inc., in 1966. He developed an inbred mutant strain with the same characteristics from this mouse and named it CTS [6]. During the course of developing the CTS strain, three lines were separated. One line that was separated at the fourth generation was bred as an inbred strain with normal eyes and was recently named NCT [7]. Two other lines were separated at the sixth generation in order to develop euglycemic and hyperglycemic strains. Although euglycemic and slightly hyperglycemic strains were bred by selective breeding for 13 generations, the development of the hyperglycemic strain ended unsuccessfully. While maintaining these two lines, we discovered a female mouse showing diabetic

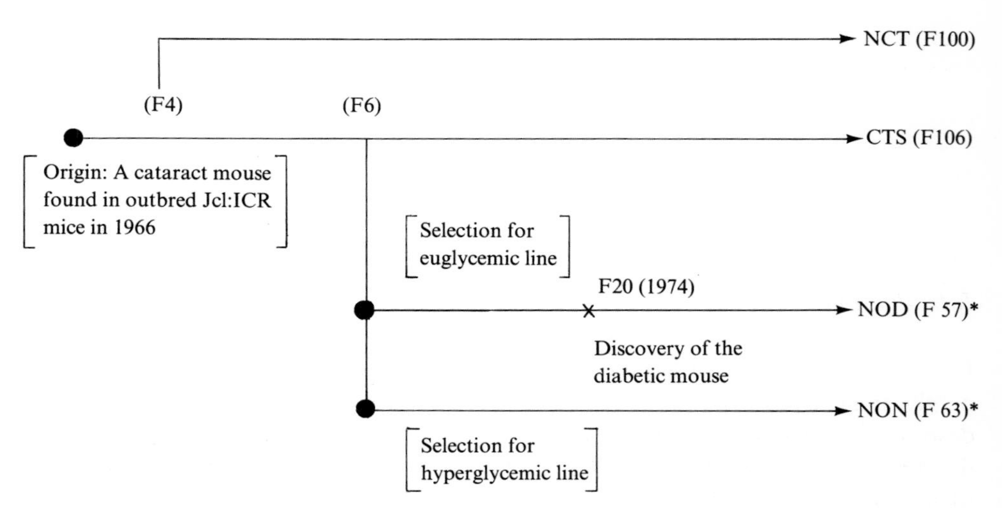

Fig. 1. Genealogy of the NON mouse and its related strains.
*: Generation after discovery.

signs in the former euglycemic line and established an inbred NOD strain from the offspring of the diabetic female mouse in 1980. However, we could not obtain a control strain for the NOD strain from the offspring of the diabetic female mouse. Therefore, we bred an inbred strain as a control for the NOD strain and named it NON [1].

Genetic characteristics of the NON mouse and its related strains

The genetic characteristics of the NON mouse and its related strains were examined through the use of three types of genetic markers: morphological traits (coat color), biochemical markers (isozymes and nonenzyme proteins) and immunological markers.

First, we determined the coat color genes by the coat color phenotype of the F1 progeny from the cross of the subject strain and the II tester strain bred for testing coat color genes [8]. The agouti (a), brown (b), dilute (d), and piebald (s) loci were examined as the hidden coat color genes. As shown in Table I, heterozygosity was not observed in these coat color loci of the strains, but a difference among the strains was observed at the agouti locus. Namely, the agouti locus of the NON and NOD strains was agouti, but that of the CTS and NCT strains was nonagouti.

Secondly, we examined the 24 loci on 12 chromosomes as biochemical marker genes by electrophoretic techniques. Samples of plasma, urine, hemolysate, and

TABLE I
Coat color genes in the NON mouse and its related strains

Chr.no.	2	4	7	9	14
Locus	a	b	c	d	s
NON	A	B	c	D	S
NOD	A	B	c	D	S
CTS	a	B	c	D	S
NCT	a	B	c	D	S

stomach, liver, intestinal, kidney and pancreas homogenate from each animal were utilized for electrophoresis using starch gel, agarose gel and cellulose acetate plates. The compositions of the buffer systems and the staining mixture were as reported previously [9–11]. The biochemical marker loci examined were: isocitrate dehydrogenase-1 (Idh-1), peptidase-3 (Pep-3), alkaline phosphatase-1 (Akp-1), amylase-2 (Amy-2), carbonic anhydrase-2 (Car-2), alcohol dehydrogenase-3 complex (Adh-3), glucose-6-phosphate dehydrogenase-1 (Gpd-1), major urinary protein (Mup-1), aldehyde dehydrogenase-1 (Ahd-1), phosphoglucomutase-1 (Pgm-1), lactate dehydrogenase regulator-1 (Ldr-1), γ-glutamyl cyclotransferase (Ggc), hemoglobin β-chain (Hbb), glucosephosphate isomerase-1 (Gpi-1), esterase-1 (Es-1), esterase-2 (Es-2), glutamate oxaloacetate transaminase-2 (Got-2), leucine arylaminopeptidase-1 (Lap-1), malic enzyme, supernatant (Mod-1), transferrin (Trf), esterase-3 (Es-3), esterase-10 (Es-10), glutamic-pyruvic transaminase-1 (Gpt-1), and aldehyde dehydrogenase-2 (Ahd-2). No genetic variance within each strain was recognized in these biochemical marker genes (Table II). However, of the 24 markers examined, 10 markers showed phenotypic differences among the strains. They were Idh-1, Amy-2, Car-2, Ahd-1, Pgm-1, Gpi-1, Lap-1, Mod-1, Es-10 and Gpt-1.

Thirdly, we analyzed the six loci on four chromosomes as immunological marker genes by the complement-dependent microcytotoxicity test using a flat type titration plate [12]. Monoclonal antibodies against Ly and Thy-1 antigens were purchased from the Meiji Institute of Health Science (Tokyo, Japan). Conventional antibodies for the study of the class I MHC genes were provided by Dr. H. Katoh, the Central Institute for Experimental Animals. Table III shows the results for the immunological marker genes. There was also no heterozygous genotype within each strain, but of the six markers examined, five markers showed genetic differences among the strains. They were Ly-2, Ly-3, Thy-1 and H-2 K and D regions.

These results on the genetic markers indicate that the NON mouse and its related strains have high genetic uniformity but are mutually independent inbred strains.

TABLE II
Biochemical marker genes in the NON mouse and its related strains

Chr.no.	1	1	1	3	3	3	4	4	4	5	6	6	7	7	8	8	8	9	9	9	11	14	15	19
Locus	Idh-1	Pep-3	Akp-1	Amy-2	Car-2	Adh-3	Gpd-1	Mup-1	Ahd-1	Pgm-1	Ldr-1	Ggc	Hbb	Gpi-1	Es-1	Es-2	Got-2	Lap-1	Mod-1	Trf	Es-3	Es-10	Gpt-1	Ahd-2
NON	b	b	b	a	b	a	b	a	a	a	a	a	s	a	b	b	b	a	a	b	c	a	a	a
NOD	a	b	b	b	a	a	b	a	a	a	a	a	s	a	b	b	b	b	b	b	c	a	b	a
CTS	b	b	b	a	b	a	b	a	b	b	a	a	s	b	b	b	b	b	a	b	c	a	a	a
NCT	a	b	b	a	a	a	b	a	a	a	a	a	s	a	b	b	b	a	a	b	c	b	a	a
C57Bl/6J	a	a	a	a	a	b	a	b	a	a	a	b	s	b	a	b	b	a	b	b	a	a	a	a

TABLE III
Immunological marker genes in the NON mouse and its related strains

Chr.no.	6	6	9	17	17	19
Locus	Ly-2	Ly-3	Thy-1	H-2K	H-2D	Ly-1
NON	2	2	1	b	non	2
NOD	1	1	2	d	b	2
CTS	1	1	1	cts	cts	2
NCT	1	1	1	s	nct	2
B6	2	2	2	b	b	2

non: unique MHC for the NON mouse
cts: unique MHC for the CTS mouse
nct: unique MHC for the NCT mouse

Conclusions

It remains unknown as to whether or not genetic mechanisms are involved in the glucose intolerance characteristics and unique renal lesions of the NON mouse. Recently, we have demonstrated that the renal lesions of the NON mouse include PAS, lipid and IgM positive glomerular intracapillary deposits [7]. Harada et al. [7, 13, 14] also have revealed that the CTS mouse shows T-lymphocytopenia and decreased T cell-mediated immunity. Therefore, the NON and CTS mice may carry a genetic immune disorder. Interestingly, genetic polymorphisms are observed not only in the biochemical markers but also in the immunological markers among the NON mouse and its related strains. If a common genetic background is related to the development of immune disorder in the NON, CTS and NOD mice, the NON mouse and its related strains may be useful for further study of the genetic mechanisms of immune disorders.

References

1 Makino S, Kunimoto K, Muraoka Y, Mizushima Y, Katagiri K, Tochino Y. Breeding of a non-obese, diabetic strain of mice. Exp Anim 1980; 29: 1–13.

2 Makino S, Hayashi Y, Muraoka Y, Tochino Y. Establishment of the nonobese-diabetic (NOD) mouse. In: Sakamoto N, Min HK, Baba S, eds. Current topics in clinical and experimental aspects of diabetes mellitus. Amsterdam: Elsevier, 1985; 25–32.

3 Ohgaku S, Morioka H, Sawa T, Yano S, Yamamoto H, Okamoto H, Tochino Y. Reduced expression and restriction fragment length polymorphism of insulin gene in NON mice, a new animal model for nonobese, noninsulin-dependent diabetes. In: Shafrir E, Renold AE, eds. Frontiers in diabetes research. Lessons from animal diabetes II. London: John Libbey, 1988; 319–323.

4 Kano Y. Insulin secretion in NOD (non-obese diabetic) and NON (non-obese non-diabetic) mouse. J Kyoto Pref Univ Med 1988;97:295–308.

5 Watanabe Y, Yoshida F, Koho N, Itoho Y, Fukatsu A, Matsuo S, Hotta N, Sakamoto N. Renal histopathological findings of the NON/Shi mouse. In: Soc Diabetes Anim, Ed. Osaka: Medical Journal Co., 1989;246–253 (in Japanese).

6 Ohotori H, Yoshida T, Inuta T. 'Small eye and cataract', a new dominant mutation in the mouse. Exp Anim 1968;17:91–96 (in Japanese).

7 Makino S, Muraoka Y, Harada M, Kishimoto Y, Takao K. Characteristics of the NOD mouse and its relatives. In: Larkins R, Zimmet P, Chisholm D, eds. New lessons from diabetes in animals. Amsterdam: Elsevier, 1989;747–750.

8 Kondo K, Esaki K. Breeding of tester strains for coat colour genes. Exp Anim 1961;11:194–196 (in Japanese).

9 Nagase S, Takahashi S, Esaki K, Mizuoka G. Methods for the detection of biochemical marker genes in inbred strains of mouse. In: Jpn Soc Immunol, Ed. Methods of immunological experiment (VIII). Kanazawa: Maeda Printing Co., 1979;2341–2356 (in Japanese).

10 Eicher EM, Womack JE. Chromosomal location of soluble glutamic-pyruvic transaminase-1 (Gpt-1) in the mouse. Biochem Genet 1977;15:1–8.

11 Mather PB, Holmes RS. Biochemical genetics of aldehyde dehydrogenase isozymes in the mouse: Evidence for stomach- and testis-specific isozymes. Biochem Genet 1984;22:981–995.

12 Shiroishi T, Sagai T, Moriwaki K. A simplified micro-method for cytotoxity testing using a flat-type titration plate for the detection of H-2 antigens. Microbiol Immunol 1981;25:1327–1334.
13 Yagi H, Suzuki S, Matsumoto M, Makino S, Harada M. Immune deficiency of the CTS mouse. I. Deficiency of in vitro T cell-mediated immune response. Immunol Invest 1990;19:279–295.
14 Yagi H, Nagata M, Takeuchi M, Watanabe A, Arimura A, Hashimoto S, Makino S, Harada M. Immune deficiency of cataract Shionogi (CTS) mouse. II. Impaired in vivo T cell-mediated immune response. Immunol Invest 1990;19:493–505.

CHAPTER 2

Biological characteristics of the NOD mouse, a related strain of the NON mouse

MINORU HARADA[a] and SUSUMU MAKINO[b]

[a]*Shionogi Research Laboratories, Shionogi & Co., Ltd., Sagisu, Fukushimaku, Osaka, Japan*
[b]*Shionogi Aburahi Laboratories, Shionogi & Co., Ltd., 1405, Gotanda, Koka-cho, Koka-gun, Shiga, Japan*

Current concepts of a new animal model: The NON mouse
Edited by N. Sakamoto, N. Hotta and K. Uchida

Contents

Introduction

The NON (non-obese non-diabetic) mouse together with the CTS (cataract Shionogi) mouse has been established from a closed colony of Jcl : ICR mice by Makino and co-workers as inbred sister strains of the NOD (non-obese diabetic) mouse which is now widely used as a good model for Type I diabetes mellitus. The pedigree and mutual relationship between these strains are described in detail in this monograph. As a reference for reviewing the biological characteristics of the NON mouse, the profile of the NOD mouse is shown in this article.

Pathological characteristics

The NOD mouse develops spontaneous diabetes characterized by polyuria, polydipsia, hyperglycemia, glucosuria, hypercholesteremia and rapid weight loss, and dies within 1 to 2 months after onset of diabetes [1]. When histologically examined, insulitis, i.e., mononuclear cell infiltration into pancreatic islets, begins to occur in both sexes at around 5 weeks of age and thereafter increases gradually in frequency, the incidence reaching more than 90% at 9 weeks of age (Table I and Fig. 1). In contrast, the development of overt diabetes is markedly different depending on the sex (Fig. 2). In females, overt diabetes can first be observed at around 13 weeks of age and then increases gradually, the cumulative incidence up to 30 weeks of age being approximately 80%. On the other hand, the occurrence of diabetes is restricted to a small number of males, the cumulative incidence up to 30 weeks being less than 20%. Overt diabetes occurs suddenly without any predictable symptoms. Pancreatic islets of the diabetic mice are destroyed or markedly atrophied if there are any, and typical insulitis can no longer be observed. Therefore, β cells which secrete insulin cannot be detected by means of AF (aldehyde-fuchsin) staining or immunohistochemical staining using peroxidase-labeled anti-insulin antibody. The majority of the mononuclear cells infiltrating the islets are T-lymphocytes [2]. A significant number of B lymphocytes are located around the islets or near the ducts but not in the islets. Infiltration of asialoGM$^+$ CD8 cells (NK cells?) is controversial. Of the T lymphocytes, CD4$^+$ T lymphocytes are always predominant, but a considerably high number of

TABLE I
Incidence of lymphocytic infiltration in the pancreatic islets with age in the NOD and its relative strains. The number in parentheses indicates the number of mice.

Sex	Age (wks)	NOD	NON	CTS
Female	3	0% (25)	0% (9)	0% (9)
	5	52% (25)	0% (14)	0% (10)
	9	92% (25)	8% (13)	0% (10)
	⩾ 20	92% (25)	14% (14)	0% (19)
Male	3	0% (25)	0% (16)	0% (10)
	5	36% (25)	0% (14)	0% (10)
	9	92% (25)	7% (15)	0% (10)
	⩾ 20	94% (25)	0% (8)	0% (10)

$CD8^+$ T lymphocytes can be found in the pancreata of newly diagnosed diabetic mice.

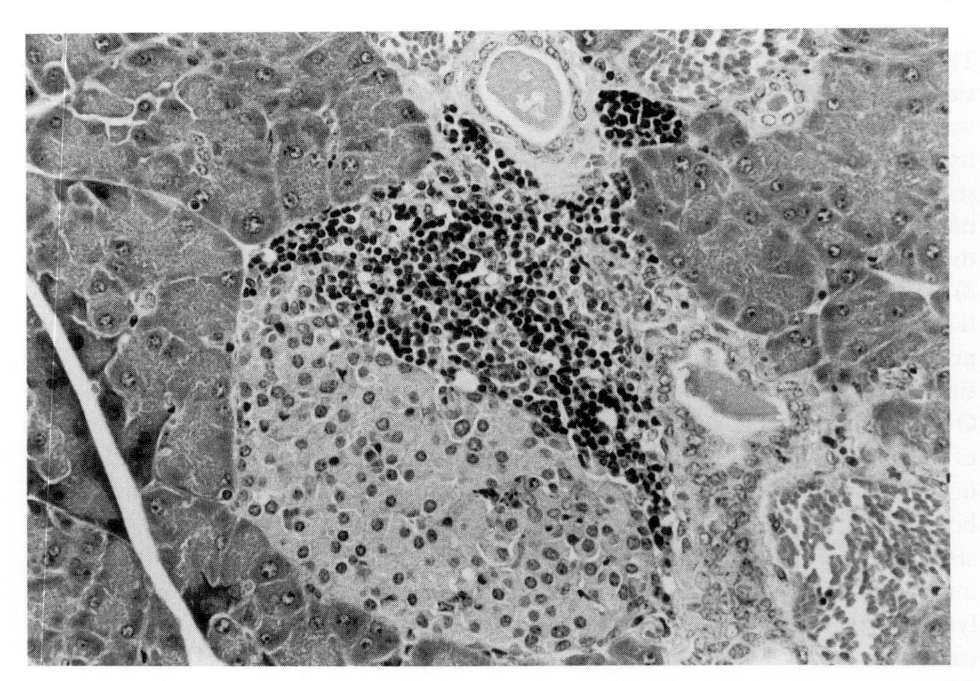

Fig. 1. Mononuclear cell infiltration (insulitis) into a pancreatic islet of a NOD mouse

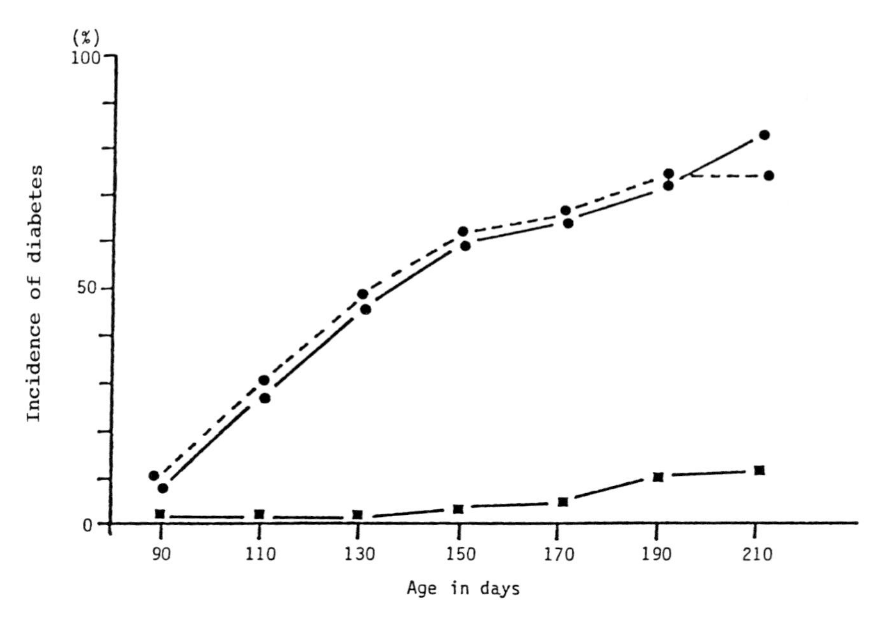

Fig. 2. Cumulative incidence of the spontaneous onset of overt diabetes in NOD mice. ●——●: Breeder females ($n=43$) ■——■: Males ($n=45$) ●----●: Virgin females ($n=36$) Ref. [1].

Genetic regulation of pathogenesis

As mentioned above, insulitis occurs in almost all mice of both sexes, while overt diabetes occurs in a large number of females but only in a small number of males. This suggests that insulitis is a more fundamental pathological change than the β-cell destruction leading to overt diabetes. On this line, Makino et al. [3] studied the mode of inheritance of the development of insulitis by crossing NOD mice with C57BL/6J mice, and concluded that two recessive genes on different autosomal chromosomes regulate the pathogenesis (Table II). However, studying the offspring of NOD and C57BL/10 mice, Wicker et al. [4] suggested that three recessive genes are involved in the production of diabetes. Hattori et al. [5] examined the correlation between the H-2 haplotype and occurrence of diabetes using back-cross and F_2 generations between NOD and C3H/He strains, and found that diabetes develops only in the H-2^{NOD} homozygous animals. Based on this result, they concluded that one of the diabetogenic genes is linked to the H-2 locus on chromosome 17. In addition, they showed that the NOD mouse lacks mRNA for the class II MHC I–E_α chain. Acha-Orbea and McDevitt [6] analyzed cDNA for class II I–A_α and I–A_β chains, and demonstrated that no difference can be seen between cDNAs for I–A_α^{NOD} and I–A_α^{d} chains, but that five bases from Nos. 248 to 252 for the I–A_β^{NOD} chain differ from

TABLE II
Incidence of insulitis in various generations between NOD and C57BL/6J(B6) mice. The number in parentheses indicates the number of mice examined. The mice were sacrificed at an age of nine weeks. Ref. [3].

Generation	Insulitis %		Relative %		(Expected*)
	Female	Male	Female	Male	Both sexes
NOD	89.7 (29)	70.4 (27)	100	100	100
C57BL(B6)	0 (30)	0 (21)	0	0	0
(NOD × B6)F1	0 (75)	0 (66)	0	0	0
F2	3.9 (155)	1.4 (148)	4.3	2.0	6.2
F1 × NOD	23.7 (93)	12.1 (107)	26.4	17.2	25.0
F1 × B6	0 (85)	0 (90)	0	0	0

* Two-recessive-gene model.

those for the I–A_β^d chain. The difference means that the two amino acids at Nos. 56 and 57 of the I–A_β^{NOD} chain are His–Ser in place of Pro–Asp of the I-A_β^d chain. This unique sequence is matched to Todd et al. [7] who showed that the amino acid of No. 57 of the HLA–DQ_β chain of Type I diabetes patients of a Caucasian population is not Asp, which can be seen in the chain of healthy persons. Paying attention to such a peculiar H-2 haplotype as summarized in Table III [8], two immunogenetic approaches have been taken to obtain further evidence of the correlation between H-2 and pathogenesis. One is a study using I–E and I–A transgenic mice and the other is a study using H-2 congenic mice.

Nishimoto et al. [9] mated NOD mice with C57BL/6J transgenic mice to which the I-E_α^d gene had been introduced and studied the incidence of insulitis in their off-

TABLE III
H-2 haplotypes of NOD and reference mice. Ref. [8]

Strain	H-2 haplotype			
	K	I–A	I–E	D
NOD	d	nod	absent	b
CTS	cts	nod	absent	cts
NON	non	non	present	b
C3H/He	k	k	k	k
BALB/c	d	d	d	d
C57BL/6J	b	b	b	b

spring. Insulitis was not observed at all at the F_1 generation. At the back-cross generation of F_1x NOD, the H-2 genotype segregates into the following four types: (i) I–A homozygous (NOD/NOD) I–E homozygous (−/−), (ii) I–A homozygous (NOD/NOD) I–E heterozygous (+/−), (iii) I–A heterozygous (NOD/b) I–E homozygous (−/−), and (iv) I–A heterozygous (NOD/b) I–E heterozygous (+/−). Insulitis was observed in approximately half the mice of type (i) and in one mouse of type (iii) (Table IV), which suggests that the I–E^d_α gene prevents the development of insulitis. This was confirmed by introducing the I–E^d_α gene directly into the fertilized eggs of NOD mice [10]. Analysis of the role of I–A using transgenic mice has also been performed in several laboratories. Miyazaki et al. [11] demonstrated that the development of insulitis can be significantly suppressed in the double transgenic NOD mice expressing I–A^k_α and I–A^k_β chains (Table V). Similar results were also obtained in other laboratories [12, 13]. From these findings, it is evident that both the absence

TABLE IV
Incidence of insulitis in the pancreatic islets of back-cross progeny between NOD and I–E^d_α transgenic C57BL/6J mice. Ref. [9]

	n	Class II phenotype	Insulitis
BC group 1	19	I–A^b(−), I–E(−)	9/19 (47%)
2	18	I–A^b(−), I–E(+)	0/18 (0%)
3	21	I–A^b(+), I–E(−)	1/21 (5%)
4	15	I–A^b(+), I–E(+)	0/15 (0%)
F_1 NOD × B6 (E^d_α)	8	I–A^b(+), I–E(+)	0/ 8 (0%)
Parent NOD	14	I–A^b(−), I–E(−)	13/14 (93%)

TABLE V
Incidence of insulitis in I-A^k transgenic NOD mice. Ref. [11]

Transgenic mice	Sex	No. of mice	Insulitis (%)
NOD	F	16	13/16 (81%)
	M	14	12/14 (85%)
NOD-A^k_α	F	4	3/4 (75%)
	M	4	3/4 (75%)
NOD-A^k_β	F	24	20/24 (83%)
	M	21	18/21 (86%)
NOD-$A^k_\alpha A^k_\beta$	F	15	5/14 (36%)
	M	21	4/15 (27%)

of I–E and the presence of unique I–A are associated with the pathogenesis in the NOD mouse.

In the other attempt to verify the correlation between the H-2 haplotype and pathogenesis, Makino et al. [14] bred several H-2 congenic strains, and demonstrated that a considerably high incidence of insulitis can be observed in the NON mice whose H-2 was replaced with $H\text{-}2^{NOD}$ (Table VI). In contrast, C57BL/6J congenic mice having $H\text{-}2^{NOD}$ did not develop insulitis at all. Furthermore, NOD congenic mice having $H\text{-}2^{NON}$ failed to develop insulitis (Table VII). All these findings show that the H-2 region plays a crucial role in the pathogenesis.

Little is known of the other diabetogenic gene(s). However, it should be noted that the NON mouse shares the H-2-nonlinked diabetogenic gene(s). This was first suggested by the crossing experiments. In contrast to the two-gene regulation in the progeny between NOD and C57BL/6J mice described above, development of insulitis was regulated by a single recessive gene in the case of the offspring of NOD and NON mice (Table VIII). Because the H-2 haplotype of the NON mouse is different from that of the NOD mouse (Table III), the common gene(s) is (are) not linked to the H-2 locus. This was supported by the aforementioned finding that the NON congenic mice having $H\text{-}2^{NOD}$ develop insulitis.

TABLE VI

Incidence of insulitis in NON congenic mice carrying the MHC of NOD mice (N3F1). The mice were sacrificed at nine weeks of age. Ref. [14]

Sex	Genotypes of the class I MHC K region		
	(NOD/NOD)	(NOD/NON)	(NON/NON)
Female	14/23	2/13	0/9
Male	16/24	0/18	0/5
Total (%)	30/47 (63.8)	2/31 (6.5)	0/14 (0)

TABLE VII

Incidence of insulitis in NOD congenic mice carrying the MHC of NON mice (N6F1). The mice were sacrificed at nine weeks of age. Ref. [14]

Sex	Genotypes of the class I MHC K region		
	(NOD/NOD)	(NOD/NON)	(NON/NON)
Female	4/5	4/13	0/15
Male	3/3	3/12	0/8
Total (%)	7/8 (87.5)	7/25 (28.0)	0/23 (0)

TABLE VIII
Incidence of insulitis in various generations between NOD and NON mice. The number in parentheses indicates the number of mice examined. The mice were sacrificed at nine weeks of age. Ref. [14]

Generation	Insulitis %		Relative %		
	Female	Male	Female	Male	(Expected*) Both sexes
NOD	89.7 (29)	70.4 (27)	100	100	100
NON	0 (15)	0 (14)	0	0	0
(NOD × NON)F1	1.6 (124)	1.5 (131)	1.8	2.1	0
F2	29.7 (91)	22.8 (79)	33.1	32.4	25.0
F1 × NOD	39.0 (200)	41.6 (209)	43.5	59.1	50.0
F1 × NON	0.9 (113)	0 (91)	1.0	0	0

* One-recessive-gene model.

Immunological mechanism of pathogenesis

Ample evidence has elucidated the crucial role of T lymphocytes in the pathogenesis. This was first suggested by the potent suppression of overt diabetes by treatments with anti-thymocyte serum or anti-Thy1.2 antibodies [15]. Definite evidence was provided by the finding that insulitis, as well as overt diabetes, was completely abolished in athymic nude mice with NOD background (Table IX) [16]. Neonatal thymectomy also suppressed the development of insulitis [17], although less potently as compared

TABLE IX
Absence of insulitis and overt diabetes in the NOD athymic nude mice. Ref. [21]

Age (wks)	Sex	Urinary glucose	Blood glucose*	Insulitis
15	M	Neg.	156	Neg.
17	F	Neg.	117	Neg.
21	M	Neg.	101	Neg.
28	F	Neg.	107	Neg.
	F	Neg.	109	Neg.
	M	Neg.	122	Neg.
47	M	Neg.	123	Neg.
	F	Neg.	191	Neg.
56	F	Neg.	125	Neg.
60	F	Neg.	76	Neg.

* In units of mg/dl

with congenital thymectomy by introducing the athymic nude gene. Further evidence came from the cell transfer experiments. Neonates [18], thymectomized and X-ray-irradiated adults [19] or B lymphocyte-reconstituted adults after thymectomy and irradiation [20] were usually employed as the recipients. Most of these studies showed that CD4^{+} T lymphocytes play the main role in the development of insulitis, and that both CD4^{+} and CD8^{+} T lymphocytes are necessary to produce diabetes. However, the possibility of the participation of the host-derived T lymphocytes, such as the T lymphocytes generated after birth, and radiation-resistant T lymphocytes cannot be ruled out in these experiments, since insulitis and/or overt diabetes did occur in some control recipients with no transferred T lymphocytes. As compared with these recipients, athymic nude congenic mice can be better recipients, because neither diabetes nor insulitis can be observed at all at least up to 60 weeks of age (Table IX) [21]. From our study of cell transfer to the athymic nude congenic mice, it was revealed that CD8^{+} T lymphocytes alone were incapble of inducing any pathological changes. On the other hand, CD4^{+} T lymphocytes alone were capable of causing insulitis but not overt diabetes. Transfer of whole spleen cells or simultaneous transfer of purified CD4^{+} and CD8^{+} T lymphocyte subsets produced both insulitis and diabetes. In addition, diabetes was induced when CD8^{+} T lymphocytes were transferred to the nude recipients in which insulitis had been established by the transfer of CD4^{+} T lymphocytes alone (Table X) [21]. These results suggest that CD8^{+} T lymphocytes activated by the helper action of CD4^{+} T lymphocytes play the major role in the β-cell destruction. The other approach is the isolation and characterization of the islet-infitrating T lymphocytes. Haskins et al. [22] established a

TABLE X

Production of insultis and overt diabetes in NOD nude mice by sequential transfer of CD4^{+} and CD8^{+} T lymphocytes. Ref. [21]

Cells transferred	CD^{+} T lymphocytes	CD4^{+} and CD8^{+} T lymphocytes
No. of recipients	3(F) + 4(M)	6(F) + 6(M)
Incidence of Insulitis	5/7	10/12
Diabetes	0/7	10/12

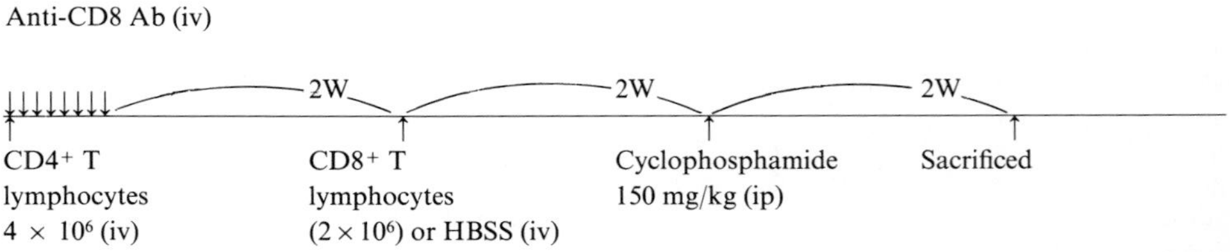

Anti-CD8 Ab (iv)

CD4^{+} T lymphocytes 4×10^{6} (iv)	→ 2W →	CD8^{+} T lymphocytes (2×10^{6}) or HBSS (iv)	→ 2W →	Cyclophosphamide 150 mg/kg (ip)	→ 2W →	Sacrificed

$CD4^+$ T cell clone from the islet-infiltrating lymphocytes, and showed that the cloned T lymphocytes were cytotoxic to β cells. Very recently, they found that transfer of these cloned cells to newborn NOD mice induces diabetes [23]. These findings suggest that some $CD4^+$ T lymphocytes are capable of inducing diabetes, as well as insulitis, by themselves, unless the host-derived T lymphocytes are co-operative.

In contrast to the role of T lymphocyte-mediated immunity, no definite evidence for the involvement of humoral immunity has been obtained. Hari et al. [24] succeeded in producing monoclonal antibodies to an islet surface antigen by fusing the NOD spleen cells with myeloma cells, and they showed that this monoclonal antibody is of the IgG_1 isotype and impairs β cells through ADCC but not through the complement-dependent cytotoxicity. However, no information is available as to whether a sufficient amount of the antibody for ADCC is present in the serum of the NOD mouse or not. The islet cell antibody (ICA) and anti-insulin antibody were reported by Hanafusa et al. [25], Reddy et al, [26] and Maruyama et al. [27]. However, the time course of the appearance of these antibodies does not parallel that of the onset of insulitis and diabetes, and therefore their etiological meaning is unclear.

Manipulation of pathogenesis

Considering that the development of insulitis and overt diabetes depends on the autoimmune mechanism, a number of attempts have been made to manipulate the pathogenesis by means of immunological methods. First, treatments with antibodies to T lymphocytes, such as anti-thymocyte serum [15], anti-Thy1.2 antibody [15], anti-CD4 antibody [28], anti-CD8 antibody [29], anti-Vβ8 T cell receptor antibody [30] (and anti-I–A antibody [31]), are known to suppress insulitis and/or diabetes. Secondly, some immunosuppressants such as cyclosporin [32] and FK506 [33] are also suppressive. Third, some kinds of cytokine, e.g., interleukin 1 and tumor necrosis factor (TNF_α), and a TNF_α-inducer, OK432, are reported to prevent the diabetes [34]. With these agents, multiple injections are required to obtain definite suppression. In contrast, a single injection of live BCG [35] or Freund's complete adjuvant [36] produces potent and long-lasting suppression. The transgene therapy described above is also a means of immunological manipulation. As non-immunological methods, infection with LCMV (lymphocyte choriomeningitis virus) [37] and MHV (mouse hepatitis virus) [38] reduces the incidence of diabetes. Also, the change in protein content of a diet is known to have a great effect on the occurrence of diabetes [39]. Nicotinamide, an inhibitor of poly (ADP-ribose) synthetase and the other ADP-ribosyl transferase, suppresses the development of diabetes as long as it is continuously administered [40]. On the other hand, cyclophosphamide potentiates the pathogenesis [41]. This agent accelerates the infiltration of mononuclear cells into pancreatic islets and promotes the onset of diabetes in both sexes. Such an effect is particularly obvious

TABLE XI
Accelerated onset of overt diabetes in NOD mice by cyclophosphamide. Ref. [41]

Dose of cyclo-phosphamide	Age at the first injection	Sex	Incidence of overt diabetes* Positive/Total (%)
150 mg/kg	5 wks	F	5/11 (46%)
		M	8/19 (42%)
	8–9 wks	F	14/22 (64%)
		M	16/22 (72%)
	12–14 wks	F	15/19 (79%)
100 mg/kg	12–14 wks	F	13/21 (64%)
75 mg/kg	12–13 wks	F	7/11 (64%)
50 mg/kg	12–13 wks	F	7/18 (39%)
Saline	5 wks	F	0/18
		M	0/20
	12–13 wks	F	0/10

* Incidence two weeks after two injections of cyclophosphamide two weeks apart.

when injected twice at a two-week interval (Table XI). Because it does not induce any pathological changes in the other mouse strains [41] or in NOD athymic nude mice [42], the effect cannot be attributed to its toxic action against β cells, but is probably due to the elimination of suppressor T lymphocytes [43].

Closing remarks

The NOD mouse is regarded as a good model for Type I diabetes mellitus and a tissue-restricted autoimmune disease. Therefore, substantial data have been accumulated regarding the pathogenic mechanism, as summarized in the present review. However, several problems, such as identification of the antigens and possible participation of retrovirus infection, have not yet been fully analyzed. Furthermore, although a variety of attempts have been made to prevent the pathogenesis, no satisfactory prophylactic or therapeutic method applicable to humans has yet been developed. To settle these problems, further investigation of the NOD mouse is necessary. These trials will also be meaningful as a reference for the study of biological characteristics of related strains such as the NON and CTS mice.

References

1 Makino S, Kunimoto K, Muraoka Y, Mizushima Y, Katagiri K, Tochino Y. Breeding of a non-obese diabetic strain of mice. Exp Anim 1980;29:1–13.

2 Miyazaki A, Hanafusa T, Yamada K, Miyagawa J, Fujino-Kurihara H, Nakajima H, Nonaka K, Tarui S. Predominance of T lymphocytes in pancreatic islets and spleen of prediabetic non-obese diabetic (NOD) mice: a longitudinal study. Clin Exp Immunol 1985;60:622–630.

3 Makino S, Muraoka Y, Kishimoto Y, Hayashi Y. Genetic analysis for insulitis in NOD mice. Exp Anim 1985;34:425–432.

4 Wicker LS, Miller JB, Coker LZ, McNally SE, Scott S, Mullen Y, Appel M. Genetic control of diabetes and insulitis in the nonobese diabetic (NOD) mice. J Exp Med 1987;165:1639–1654.

5 Hattori M, Buse JB, Jackson RA, Glimcher L, Dorf ME, Minami M, Makino S, Moriwaki K, Kuzuya H, Imura H, Strauss WM, Seidman JG, Eisenbarth GS. The NOD mouse: recessive diabetogenic gene in the major histocompatibility complex. Science 1986;231:733–735.

6 Acha-Orbea D, McDevitt HO. The first external domain of the nonobese diabetic mouse class II I–A_β chain is unique. Proc Natl Acad Sci USA 1987;84:2435–2439.

7 Todd JA, Bell JI, McDevitt HO. HLA-DQ gene contributes to susceptibility and resistance to insulin-dependent diabetes mellitus. Nature 1987;329:599–604.

8 Ikegami H, Makino S, Hattori M. Genetic analysis of the non-obese diabetic (NOD) mouse. Frontiers of Mucosal Immunol 1990;2:429–432.

9 Nishimoto H, Kikutani H, Yamamura K, Kishimoto T. Prevention of autoimmune insulitis by expression of I–E molecules in NOD mice. Nature 1987;328:432–434.

10 Uehira M, Uno M, Kürner T, Kikutani H, Mori K, Inomoto T, Uede T, Miyazaki J, Nishimoto H, Kishimoto T, Yamamura K. Development of autoimmune insulitis is prevented in E^d_α but not in A^k_β NOD transgenic mice. Intern Immunol 1989;1:209–213.

11 Miyazaki T, Uno M, Uehira M, Kikutani H, Kishimoto T, Kimoto M, Nishimoto H, Miyazaki J, Yamamura K. Direct evidence for the contribution of the unique I–A^{NOD} to the development of insulitis in non-obese diabetic mice. Nature 1990;345:722–724.

12 Lund T, O'Reilly L, Hutching P, Kanagawa O, Simpson E, Graavely R, Chandler P, Dyson J, Picard JK, Edwards A, Kioussis D, Cooke A. Prevention of insulin-dependent diabetes mellitus in non-obese diabetic mice by transgenes encoding modified I–A_β-chain or normal I–E_α-chain. Nature 1990;345:727–729.

13 Slattery RM, Kjer-Nielsen L, Allison J, Charlton B, Mandel T, Miller JFAP. Prevention of diabetes in non-obese diabetic I–A^k transgenic mice. Nature 1990;345:724–726.

14 Makino S, Muraoka Y, Harada M, Hayashi Y. Characteristics of the NOD mouse and its relatives. New Lessons from Diabetes in Animals in Diabetes. Amsterdam: Elsevier, 1988;747–750.

15 Harada M, Makino S. Suppression of overt diabetes in NOD mice by anti-thymocyte serum or anti-Thy1.2 antibody. Exp Anim 1986;35:501–504.

16 Makino S, Harada M, Kishimoto Y, Hayashi Y. Absence of insulitis and overt diabetes in athymic nude mice with NOD genetic background. Exp Anim 1986;35:495–498.

17 Ogawa M, Maruyama T, Hasegawa T, Kanaya T, Kobayashi F, Tochino Y, Uda H. The inhibitory effectof neonatal thymectomy on the incidence of insulitis in non-obese diabetes (NOD) mice. Biomed Res 1984;6:103–105.

18 Bendelac A, Carnaud C, Biotard C, Bach JF. Syngeneic transfer of autoimmune diabetes from diabetic NOD mice to healthy neonates. J Exp Med 1987;166:823–832.

19 Miller BJ, Appel MC, O'Neill JJ, Wicker LS. Both the Lyt2+ and L3T4+ T cell subsets are required for the transfer of diabetes in nonobese diabetic mice. J Immunol 1988;140:52–58.

20 Hanafusa T, Sugihara T, Fujino-Kurihara H, Miyagawa J, Miyazaki A, Yoshida T, Yamada K, Nakajima H, Asakawa H, Kono N, Fujiwara H, Hamaoka T, Tarui S. Induction of insulitis by adoptive

transfer with L3T4+Lyt2− T-lymphocytes in T-lymphocyte-depleted NOD mice. Diabetes 1988;37:204–208.

21 Harada M, Yagi H, Matsumoto M, Makino S. In: Analysis of the roles of L3T4+ and Lyt2+ T-cells in the pathogenesis of NOD mice by means of cell transfer to athymic nude congenic mice. Frontiers in Diabetes Research. Lessons from Animal Diabetes III, IBB, 1990; E. Shafrir, ed., 59–62.

22 Haskins K, Portas M, Bradley B, Wegmann D, Lafferty K. T-lymphocyte clone specific for pancreatic islet antigen. Diabetes 1988;37:1444–1448.

23 Haskins K, McDuffie M. Acceleration of diabetes in young NOD mice with a CD4+ islet-specific T cell clone. Science 1990;249:1433–1436.

24 Hari J, Yokono K, Yonezawa K, Amano K, Yaso S, Shii K, Imamura Y, Baba S. Immunochemical characterization of anti-islet cell surface monoclonal antibody from nonobese diabetic mice. Diabetes 1986;35:517–522.

25 Hanafusa T, Miyazaki A, Yamada K, Miyagawa J, Fujino-Kurihara H, Nakajima H, Kono N, Nonaka K, Tarui S. Autoantibodies to islet cells and multiple organs in the NOD mouse. Diabetes Nutr Metab 1988;1:273–276.

26 Reddy S, Bibby NJ, Elliott RB. Islet cell antibodies and insulin autoantibodies as predictive markers in the non-obese diabetic mouse. Diabetologia 1988; 31: 322–328.

27 Maruyama T, Takei I, Asaba Y, Yanagawa T, Takahashi T, Itoh H, Suzuki Y, Kataoka K, Saruta T, Ishii T. Insulin autoantibodies in mouse models of insulin-dependent diabetes. Diabetes Res 1989;11:61–65.

28 Koike T, Itoh Y, Ishii T, Ito I, Takabayashi K, Maruyama N, Tomioka H, Yoshida S. Preventive effect of monoclonal anti-L3T4 antibody on development of diabetes in NOD mice. Diabetes 1987;36:539–541.

29 Charlton B, Bacelj A, Mandel TE. Administration of silica particles or anti-Lyt 2 antibody prevents β-cell destruction in NOD mice given cyclophosphamide. Diabetes 1988;37:930–935.

30 Bacelj A, Charlton B, Mandel T. Prevention of cyclophosphamide-induced diabetes by anti-Vβ8 T-lymphocyte-receptor monoclonal antibody therapy in NOD/Wehi mice. Diabetes 1989;38:1492–1495.

31 Boitard C, Bendelac A, Richard MF, Carnaud C, Bach JF. Prevention of diabetes in nonobese diabetic mice by anti-I–A monoclonal antibodies: Transfer of protection by splenic T cells. Proc Natl Acd Sci USA 1988;85:9719–9723.

32 Mori Y, Suko M, Okudaira H, Matsuba I, Tsuruoka A, Sasaki A, Yokoyama H, Tanase T, Shiba T, Nishimura M, Terada E, Ikeda Y. Preventive effects of cyclosporin on diabetes in NOD mice. Diabetologia 1986;29:244–247.

33 Miyagawa J, Yamamoto K, Hanafusa T, Itoh N, Nakagawa C, Otsuka A, Katsura H, Yamagata K, Miyazaki A, Kono N, Tarui S. Preventive effect of a new immunosuppressant FK506 on insulitis and diabetes in non-obese diabetic mice. Diabetologia 1990;33:503–505.

34 Toyota T, Satoh J, Oya K, Shintani S, Okano T. Streptococcal preparation (OK-432) inhibits development of Type I diabetes in NOD mice. Diabetes 1986;35:496–499.

35 Harada M, Kishimoto Y, Makino S. Prevention of overt diabetes and insulitis in NOD mice by a single BCG vaccination. Diab Res Clin Pract 1990;8:85–89.

36 Sadelain MWJ, Qin H-Y, Lauzon J, Singh B. Prevention of Type I diabetes in NOD mice by adjuvant immunotherapy. Diabetes 1990;39:583–589.

37 Oldstone MBA. Prevention of Type I diabetes in nonobese diabetic mice by virus infection. Science 1988;239:500–502.

38 Wilberz S, Prtke HJ, Dagnaes-Hansen F, Herberg L. Persistent MHV (mouse hepatitis virus) infection reduces the incidence of diabetes mellitus in non-obese diabetic mice. Diabetologia 1991;34:2–5.

39 Elliott RB, Reddy S, Bibby NJ, Kida K. Dietary prevention of diabetes in the non-obese diabetic mouse. Diabetologia 1988;31:62–64.

40 Nakajima J, Fujino-Kurihara H, Hanafusa T, Yamada K, Miyazaki A, Migawa J, Nonaka K, Tarui

S, Tochino Y. Nicotinamide prevents the development of cyclophosphamide-induced diabetes mellitus in male non-obese diabetic (NOD) mice. Biomed Res 1985;6:185–189.

41 Harada M, Makino S. Promotion of spontaneous diabetes in non-obese diabetes-prone mice by cyclophosphamide. Diabetologia 1984;27:604–606.

42 Harada M. Immune disturbance and pathogenesis of non-obese diabetes-prone (NOD) mice. Exp Clin Endocrinol 1987;89:251–258.

43 Yasunami R, Bach J-F. Anti-supressor effect of cyclophosphamide on the development of spontaneous diabetes in NOD mice. Eur J Immunol;18:481–484.

CHAPTER 3

Analysis of the insulin genes in the NON mouse

SEIJI OHGAKU and TASUKU SAWA

The First Department of Internal Medicine, Toyama Medical and Pharmaceutical University, 2630 Sugitani, Toyama, Japan

Current concepts of a new animal model: The NON mouse
Edited by N. Sakamoto, N. Hotta and K. Uchida

Contents

Introduction

Non-insulin-dependent-diabetes mellitus (NIDDM) is the most common type of diabetes in Japan as well as in other countries. Understanding and clarification of the etiology of NIDDM will contribute to future therapy for the prevention of diabetes.

NIDDM is well known to have a strong genetical background, and is etiologically a heterogeneous disorder with common symptoms secondary to hyperglycemia. Pancreatic β-cell failure to secrete insulin appropriately and peripheral insulin resistance are common features observed in NIDDM. It is still uncertain, however, which is primary in the pathogenesis of NIDDM. Obese NIDDM is associated with peripheral insulin resistance and hyperinsulinemia. From the presence of hyperinsulinemia, pancreatic β cells in obese NIDDM appear to work normally to compensate peripheral insulin resistance. On the other hand, non-obese NIDDM is not necessarily associated with hyperinsulinemia. Our hypothesis is that at least some diabetics with non-obese NIDDM have a primary defect in the pancreatic β cell that induces diabetes. To address this hypothesis, we looked for a spontaneous diabetic animal model suitable for studying non-obese human NIDDM.

The NON and the NOD mouse are sister strains which were separated from the Jcl-ICR mouse in Shionogi Research Laboratories, Osaka, Japan [1, 2]. The NOD mouse, established as a model of insulin-dependent diabetes mellitus (IDDM), is normoglycemic before developing diabetes. The NON mouse of the other strain was known to be slightly hyperglycemic without glycosuria. We have characterized the NON mouse as an animal model of non-obese NIDDM [3]. The NON mouse is non-obese, glucose-intolerant, and is associated with hypoinsulinemia. No destruction of islets through infiltration of mononuclear cells (insulitis) was observed in the pancreas of the NON mouse. While insulin and preproinsulin mRNA contents decreased in the pancreas, insulin resistance was not significant in adipocytes of the NON mouse. These findings suggest that glucose intolerance in the NON mouse cannot be accounted for by peripheral insulin resistance, but by decrease of insulin synthesis in pancreatic β cells through some defects after the step of transcription of the insulin gene.

In the present study, the preproinsulin genes of the NON mouse were cloned and analyzed in our laboratory [4] to determine whether any structural changes are present on the insulin gene to affect its transcription.

Materials and methods

NON mice were provided from Shionogi Research Laboratories, Osaka, Japan.

Construction of the genomic library of the NON mouse

High-molecular-weight DNA was isolated from the mouse liver as described elsewhere [5], with slight modifications. Livers were put into nylon mesh (sample pack, Eiken), gently squashed, and the remaining connective tissue in the mesh was removed. Liver cells were lysed with proteinase K and sodium lauroyl sarcosinate at 50°C. DNA was then extracted with phenol/chloroform/isoamyl alcohol (25 : 24 : 1) and subjected to dialysis. After partial digestion of high-molecular-weight DNA with MboI, 20 kbp DNA fragments were isolated by sucrose gradient ultracentrifugation, ligated with λ DASH/BamHI arms (Stratagene) and assembled into phage using an in-vitro packaging kit (Amersham).

Isolation and subcloning of the insulin gene from the genomic library of the NON mouse

Approximately 6×10^5 plaque-forming units (pfu) phages were screened using ^{32}P-labelled preproinsulin I cDNA as a probe by plaque hybridization [6]. The positive plaques were further subjected to secondary and tertiary screening, and phage DNA was then prepared by glycerol step-gradient centrifugation [7]. For restriction mapping and subcloning, phage DNA was digested with EcoRI and BamHI, applied on agarose gel electrophoresis and subjected to Southern blot hybridization. DNA fragments which hybridize rat preproinsulin I cDNA were subcloned into Bluescript KS(+) plasmid (Stratagene), and propagated in E coli MC1061. Plasmid DNA was prepared by alkaline lysis and polyethyene glycol precipitation methods.

Construction of deletion mutants and determination of the nucleotide sequence of the NON mouse insulin gene.

Deletion mutants were prepared from plasmid subclones [8]. From the mutant subclones containing the appropriate size of DNA, plasmid was again prepared for the determination of its nucleotide sequence by the dideoxy chain-termination method [9] using synthetic primers to T7 promoter, and [α–^{35}S]dATP.

Results

A 1.1 × 10^6 pfu of the NON mouse genomic library was constructed. After amplification of the library, five positive clones were obtained after a third screening of 6 × 10^5 pfu using rat preproinsulin I cDNA as a probe. These positive clones were confirmed to consist of two independent clones by Southern analysis. The rat preproinsulin I cDNA probe hybridized with a 1.4 kbp EcoRI fragment of one clone, λNins-I, and with a 2.5 kbp BamHI fragment of the other clone, λNins-II. λNins-I and λNins-II were inferred to carry the full length of putative NON mouse preproinsulin I and II genes respectively from the comparison of restriction maps of the BALB/c mouse corresponding genes [10].

For further restriction mapping and sequencing analysis, these putative preproinsulin genes were subcloned into plasmid: pNins-I with a 1.4 kbp EcoRI fragment from λNins-I, and pNins-II with a 2.5 kbp BamHI fragment from λNins-II.

The subclone pNins-I, putative NON mouse preproinsulin I gene consisted of 1384 nucleotides (Fig. 1). Comparing pNins-I with the BALB/c preproinsulin I gene, the

```
gaattctgtaataactatatagaactcttcttatatatgctcaaattttacatgctagccttcaggtacatatcttgggttgttgggtattgtagaagaa   100

tgtactacagggcttcagcccagttgaccaatgagtgggctacggggtttgtgaaaggagagatggagaaggagggaccattaagtaccttgctgcctga   200

gttctgctttcctcctccctctgagggtgagctgggatctcatctgagttaagggcccagctatcaatgggaactgtgaaacagtccaaggacatcaat   300

                                   "Enhancer core sequence"
attaggtccctaacaactgcagtttcctggggaatgatgtggaaaatgctcagccaaagctgaagaaggtctcaccttctgggacaatgtcccctgctgg   400

         "Far box" like sequence"
gaactggttcatcaggccatctggtcccttattaagactataataaccctaagactaagtagatgtgttgatgtccaatgactgctttctgcagacctag   500

                                      "Nir box" like sequence"                              CAAT box
caccaggcaagtgtttggaaactgcagcttcagcccctctggccatctgcctacccaccccacctggagaccttaatgggccaaacagcaaagtccaggg   600

                     TATA box
ggcagagaggaggtactttggactataaagctggtgggcatccagtaACCCCCAGCCCTTAGTGACCAGCTATAATCAGAGACCATCAGCAAGcaggtat   700

gtactctcctctttgggcctggctccccagccaagactccagcgactttagggagaatgtgggctcctctcttacatggatcttttgctagcctcaaccc   800

tgcctatctttCAGGTCATTGTTTCAACATGGCCCTGTTGGTGCACTTCCTACCCCTGCTGGCCCTGCTTGCCCTCTGGGAGCCCAAACCCACCCAGGCT   900
                              MetAlaLeuLeuValHisPheLeuProLeuLeuAlaLeuLeuAlaLeuTrpGluProLysProThrGlnAla

TTTGTCAAACAGCATCTTTGTGGTCCCCACCTGGTAGAGGCTCTCTACCTGGTGTGTGGGGAGCGTGGCTTCTTCTACACACCCAAGTCCCGCCGTGAAG   1000
PheValLysGlnHisLeuCysGlyProHisLeuValGluAlaLeuTyrLeuValCysGlyGluArgGlyPhePheTyrThrProLysSerArgArgGluV

TGGAGGACCCACAAGTGGAACAACTGGAGCTGGGAGGAAGCCCCGGGGACCTTCAGACCTTGGCGTTGGAGGTGGCCCGGCAGAAGCGTGGCATTGTGGA   1100
alGluAspProGlnValGluGlnLeuGluLeuGlyGlySerProGlyAspLeuGlnThrLeuAlaLeuGluValAlaArgGlnLysArgGlyIleValAs
                                                                                         polyA signal
TCAGTGCTGCACCAGCATCTGCTCCCTCTACCAGCTGGAGAACTACTGCAACTAAGGCCCACCTCGACCCGCCCCACCCCTCTGCAATGAATAAAACTTT   1200
pGlnCysCysThrSerIleCysSerLeuTyrGlnLeuGluAsnTyrCysAsn***

TGAATAAGCaccaaaaaaaagagttctataatgaatgaaaaaggattgtgtatatagacatctttttctctggcatttattctgatgttatcatactatt   1300

aaaccattcttaggttggatgattatataatcatgtatgaagcttgtgataaaacaccaggaataattcaagtatctggaattc   1384
```

Fig. 1. Nucleotide and deduced amino acid sequence of the NON mouse preproinsulin I gene. Captial letters indicate exons and lowercase letters are used for introns and 5′- and 3′-flanking sequences. The numbering of nucleotide residues is indicated at the end of each line. The deduced amino acid sequence is given below the nucleotide sequence. The stop codon is indicated by three asterisks. The sequences of putative promoter, enhancers and the poly (A) signal are underlined. Reproduced from Sawa et al. [4] by copyright permission of the Journal of Endocrinology, Ltd.

putative transcription initiating site was considered at A in position 648. Transcriptional regulatory region of about 350 bp upstream from the transcription initiation site in the 5′-flanking portion of the NON mouse preproinsulin I contained all of

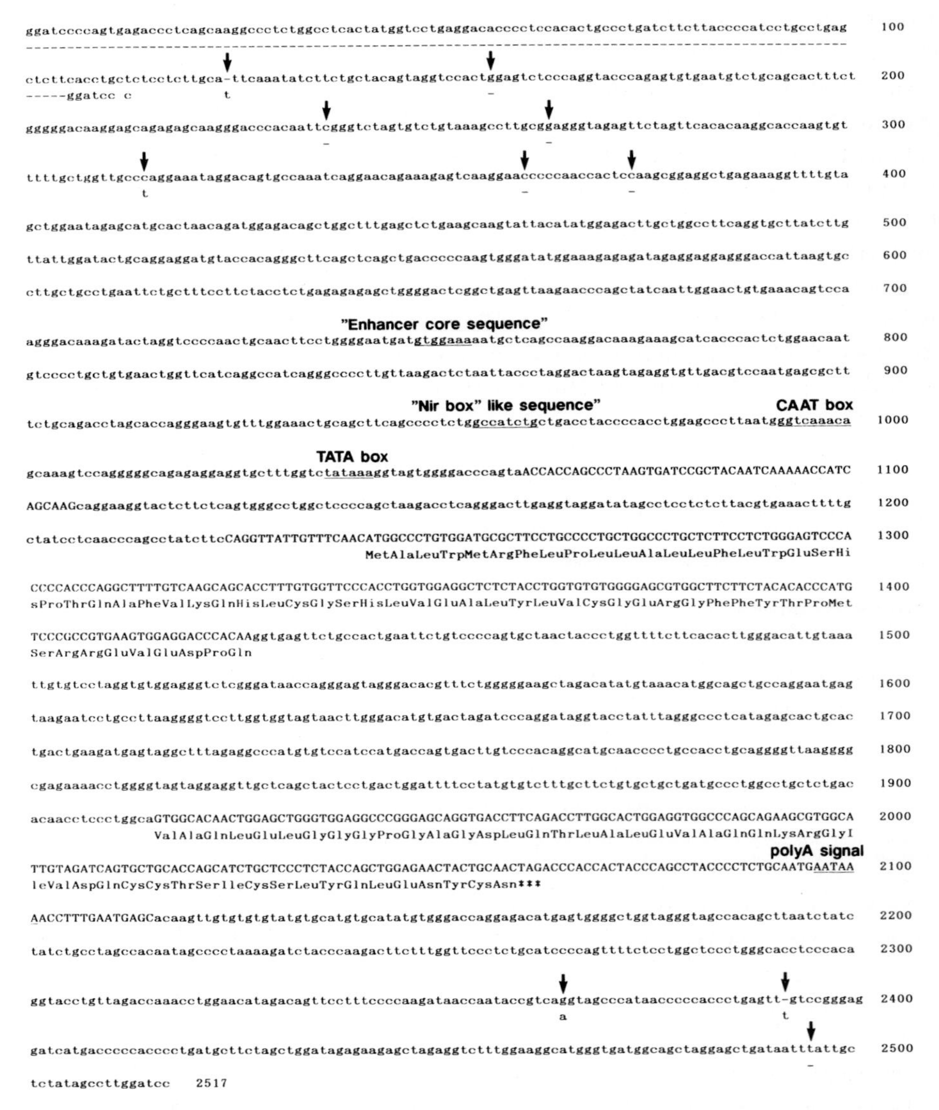

Fig. 2. Nucleotide and deduced amino acid sequence of the NON mouse preproinsulin II gene. The description system is the same as in Fig. 1. Nucleotide differences found in the BALB/c sequence are displayed beneath the NON mouse sequence and indicated by arrows. Deletions are indicated by dashes. Reproduced from Sawa et al. [4] by copyright permission of the Journal of Endocrinology, Ltd.

TATA box, CAAT box, enhancer core sequence, Nir-box-like sequence, and Far-box-like sequence. The nucleotide sequences of 5′-flanking region (positions 1–647), exon 1 (648–693), exon 2 (812–1209) including polyadenylation signal, intron (694–811) and the 3′-flanking region (1210–1384) were completely identical with those of BALB/c mouse preproinsulin I gene.

The other subclone, pNins-II that encodes the NON mouse preproinsulin II gene consisted of 2517 bp (Fig. 2), and the putative transcription initiation site was at A in position 1061. No differences in the nucleotide sequence of the transcriptional regulatory region were found between the NON and the BALB/c mouse. From the BamHI subcloning site, however, pNins-II had an additional 113 bp sequence which was absent in the BALB/c mouse gene. The subclone pNins-II also had seven point mutations in the 5′-flanking region of the preproinsulin II gene (five insertions at positions 156, 237, 264, 361, and 374; one substitution at 315; one deletion at between 124 and 125), and 3 point mutations in the 3′-flanking region (one insertion at 2495; one substitution at 2365: one deletion at 2391 or 2392). Nucleotide sequences of other regions of pNins-II, encoding the NON mouse preproinsulin II gene including exons 1 (positions 1061–1106), 2 (1225–1427), and 3 (1916–2115), and introns 1 (1107–1224) and 2 (1428–1915) were identical with those of the BALB/c mouse gene.

The deduced amino acid sequences for the NON mouse preproinsulin I and II were identical with those of BALB/c mouse preproinsulins.

Discussion

Clinically established NIDDM is characterized both by β-cell dysfunction and by peripheral insulin resistance. Evidence is accumulating [11-13] which suggests that the earliest pathologic lesion in NIDDM is peripheral insulin resistance, and hyperinsulinemia is a compensatory response for it in persons at a high risk for developing diabetes. Warram et al. [13] studied 155 normal offspring of two parents with type II diabetes. They had slower glucose metabolic rates (Kg) and higher insulin levels than control subjects. Sixteen percent of the offspring of diabetic parents developed type II diabetes during a follow-up period of 13 years. Kg at baseline among offspring who developed diabetes was significantly lower than Kg among offspring who remained non-diabetic. This reduced glucose clearance was accompanied by compensatory hyperinsulinemia. On the other hand, Temple et al. [14] claimed that hyperinsulinemia in NIDDM may not be real, from evidence in 49 patients with NIDDM, using a highly specific two-site immunoradiometric assay for insulin. From the increase of plasma level of biologically less active proinsulin and proinsulin-like molecules, they concluded that most NIDDM patients were insulin deficient in spite of hyperinsulinemia determined by usual insulin radioimmunoassays where proinsulin cross-reacts as insulin.

With regard to pancreatic β-cell function, low insulin responders have been regarded to have a higher risk of diabetes regardless of glucose-tolerance [15–17]. Non-obese Japanese people who developed NIDDM often had normal or low fasting insulin. Normo- or hypoinsulinemia with concurrent hyperglycemia indicates β-cell failure to compensate hyperglycemia. In a study of 288 Japanese subjects with impaired glucose tolerance during eight years, Kadowaki et al. [17] showed that those who have low insulin response after OGTT are at high risk of developing NIDDM [17]. To examine the hypothesis that the genetic predisposition to pancreatic β-cell failure is the primary event of NIDDM, the NON mouse as an animal model of non-obese human NIDDM was studied.

The NON mouse is a non-obese animal with glucose intolerance [1–3]. We have studied insulin binding and glucose uptake to adipocytes isolated from the NON mouse (data not shown). Insulin binding in the NON mouse was lower than in the control ICR mouse, but the difference was not significant. Scatchard analysis revealed no changes in the number of insulin receptors. Percent maximum glucose uptake to adipocytes from the basal level was lower in the NON mouse, but the decrease also failed to reach significance. These results indicate that peripheral insulin resistance is not the major factor to induce glucose intolerance in the NON mouse. Furthermore, hypoinsulinemia and decreased contents of insulin and preproinsulin mRNA in the pancreas [3] suggest dysfunction of β cells at the transcription of insulin gene.

The transcriptional regulatory region of the insulin gene is needed for its tissue-specific expression. Karlsson et al. [18] reported in the block mutagenesis study of rat insulin I gene, that four block replacement mutants spanning up to 350 bp of the 5′-flanking region immediately upstream from transcription initiation site exhibited 5 to 10 times less activity utilizing CAT assay. These mutationally sensitive regions are the TATA box, Nir box, Far box, and Enhancer core sequence.

The complete nucleotide sequences of the NON mouse preproinsulin genes were determined in our laboratory [4]. Contrary to our expectation, the NON mouse kept all these sequences mentioned above and had no structural alterations in the transcriptional regulatory region of preproinsulin I and II when compared with the corresponding genes from BALB/c mouse [10]. However, the NON mouse preproinsulin II gene contained a 113 bp additional sequence from its cloning site of 5′-flanking region (position 1–113). This 113 bp sequence which is absent in the BALB/c mouse gene may be a polymorphic region of the insulin gene as reported in humans [19, 20] and rats [21]. The relationship between NIDDM and the polymorphism in 5′-flanking region of the human insulin gene was discussed [22]. It may be also possible that the sequence in the NON mouse plays a role as a negative transcriptional regulatory ('silencer') element which is reported in rats between 2.0 and 4.0 kbp upstream of the preproinsulin I gene [23].

Linkage studies have been performed in search of the responsible genes in

NIDDM. Maturity onset of diabetes of the young (MODY) is a subgroup of NIDDM, and characterized with autosomal dominant inheritance, non-ketotic diabetes which develops at a younger age, and very few microangiopathic complications [24]. The best-studied MODY pedigree is the RW family, and the majority of its diabetic members have a reduced and delayed insulin response to glucose. Andreone et al. [25] performed a formal linkage analysis by calculating the LOD score in the family. They showed no linkage between insulin gene locus and MODY with the use of VNTR (variable-number tandem repeat) polymorphism as a genetic marker in the 5′-flanking region of the human insulin gene. Lack of linkage was also reported by Elbein et al. [26] between the insulin gene and familial NIDDM in pedigrees of Caucasian Utah Mormons. More recently, Bell et al. [27] analyzed 79 DNA markers including at least one marker on each autosome in the RW family for linkage with MODY. The analysis revealed a MODY gene to be localized on the long arm of chromosome 20 and to be tightly linked to the adenosine deaminase gene (ADA), although the identity of the MODY gene is unknown. Current evidence suggests that diabetes-susceptible genes in NIDDM may be identified through the linkage analysis, and insulin gene itself may not play a major role in the pathogenesis of NIDDM, and also suggests that the defect of genes coding trans-acting factor which regulate insulin gene transcription may be responsible for the genesis of NIDDM.

However, genetic susceptibility of diabetes is heterogeneous and insulin gene defects can not be entirely excluded in a small subset of NIDDM. The role of the additional sequence found in the NON mouse preproinsulin II gene is not known at present. Further study on insulin genes in sister strains of the NON mouse such as the ICR, CTS, and NOD mouse will provide an answer for it.

Acknowledgements

We thank Dr. H. Okamoto for the generous gift of rat preproinsulin I cDNA, and Dr. M. Yamamoto for his useful suggestions and technical advice. The study was performed in collaboration with Shionogi Research Laboratories, Osaka, Japan, and also supported, in part, by a Grant-in-Aid for Scientific Research, No. 62570507, from Ministry of Education, Science and Culture.

References

1 Makino S, Kunimoto K, Muraoka Y, Mizushima Y, Katagiri K, Tochino Y. Breeding of a non-obese, diabetic strain of mice. Exp Anim 1980;29:1–13.

2 Tochino Y. Discovery and bleeding of the NOD mouse. In: Tarui S, Tochino Y, Nonaka K, eds. Insulitis and Type I Diabetes: Lessons from the NOD Mouse. Tokyo: Academic Press, 1986;3–10.

3 Ohgaku S, Morioka H, Sawa T, et al. Reduced expression and restriction fragment length polymorphism of insulin gene in NON mice, a new animal model for nonobese, noninsulin-dependent diabetes. In: Shafrir E, Renold AE, eds. Frontiers in Diabetes Research: Lessons from Animal Diabetes II. London: John Libbey & Company Ltd, 1988;319–323.
4 Sawa T, Ohgaku S, Morioka H, Yano S. Molecular cloning and DNA sequence analysis of preproinsulin genes in the NON mouse, an animal model of human non-obese, non-insulin-dependent diabetes mellitus. J Mol Endocrinol 1990;5:61–67.
5 Cox R, Damjanov I, Abanobi SE, Sarma DSR. A method for measuring DNA damage and repair in the liver in vivo. Cancer Res 1973;33:2114–2121.
6 Benton WD, Davis RW. Screening λgt recombinant clones by hybridization to single plaques in situ. Science 1977;196:180–182.
7 Maniatis T. Molecular Cloning: A Laboratory Manual. New York: Cold Springer Harbor Laboratory, 1982.
8 Henikoff S. Unidirectional digestion with exonuclease III creates targeted breakpoints for DNA sequencing. Gene 1984; 28:351–359.
9 Sanger F, Nicklen S, Coulson AR. DNA sequencing with chain-terminating inhibitors. Proc Natl Acad Sci USA 1977;74:5463–5467.
10 Wentworth BM, Schaefer IM, Villa-Komaroff L, Chirgwin JM. Characterization of the two nonallelic genes encoding mouse preproinsulin. J Mol Evol 1986;23:305–312.
11 Lillioja S, Mott DM, Howard BV, et al. Impaired glucose tolerance as a disorder of insulin action: longitudinal and cross-sectional studies in Pima Indians. N Engl J Med 1988;318:1217–1225.
12 Haffner SM, Stern MP, Hazuda HP, Mitchell BD, Patterson JK. Increased insulin concentrations in nondiabetic offspring of diabetic parents. N Engl J Med 1988;319:1297–1301.
13 Warram JH, Martin BC, Krolewski AS, Soeldner JS, Kahn CR. Slow glucose removal rate and hyperinsulinemia precede the development of type II diabetes in the offspring of diabetic parents. Ann Intern Med 1990;113:909–915.
14 Temple RC, Carrington CA, Luzio SD, et al. Insulin deficiency in non-insulin-dependent diabetes. Lancet 1989;I:293–295.
15 Cerasi E, Luft R. 'What is inherited-what is added' hypothesis for the pathogenesis of diabetes mellitus. Diabetes 1967;16:615–627.
16 Kosaka K, Hagura R, Kuzuya T. Insulin responses in equivocal and definite diabetes, with special reference to subjects who had mild glucose intolerance but later developed definite diabetes. Diabetes 1977;26:944–952.
17 Kodowaki T, Miyake Y, Hagura R, et al. Risk factors for worsening to diabetes in subjects with impaired glucose tolerance. Diabetologia 1984;26:44–49.
18 Karlsson O, Edlund T, Moss JB, Rutter WJ, Walker MD. A mutational analysis of the insulin gene transcription control region: expression in beta cells is dependent on two related sequences within the enhancer. Proc Natl Acad Sci USA 1987;84:8819–8823.
19 Bell GI, Karam JH, Rutter WJ. Polymorphic DNA region adjacent to the 5′ end of the human insulin gene. Proc Natl Acad Sci USA 1981;78:5759–5763.
20 Bell GI, Selby MJ, Rutter WJ. The highly polymorphic region near the human insulin gene is composed of simple tandemly repeating sequences. Nature 1982;295:31–35.
21 Winter WE, Beppu H, Maclaren NK, Cooper DL, Bell GI, Wakeland EK. Restriction-fragment-length polymorphisms of 5′-flanking region of insulin I gene in BB and other rat strains: Absence of association with IDDM. Diabetes 1987;36:193–198.
22 Rotwein P, Chyn R, Chirgwin J, Cordell B, Goodman HM, Permutt MA. Polymorphism in the 5′-flanking region of the human insulin gene and its possible relation to type 2 diabetes. Science 1981;213:1117–1120.

23 Laimins L, Holmgren-Koenig M, Khoury G. Transcriptional 'silencer' element in rat repetitive sequences associated with the rat insulin I gene locus. Proc Natl Acad Sci USA 1986;83:3151–3155.
24 Fajans SS. MODY-a model for understanding the pathogenesis and natural history for type II diabetes. Horm Metab Res 1987;19:591–599.
25 Andreone T, Fajans S, Rotwein P, Skolnick M, Permutt MA. Insulin gene analysis in a family with maturity-onset diabetes of the young. Diabetes 1985;34:108–114.
26 Elbein SC, Corsetti L, Goldgar D, Skolnick M, Permutt MA. Insulin gene in familial NIDDM: lack of linkage in Utah Mormon pedigrees. Diabetes 1988;37:569–576.
27 Bell GI, Xiang K, Newman NV et al. Gene for non-insulin-dependent diabetes mellitus (maturity-onset diabetes of the young subtype) is linked to DNA polymorphism on human chromosome 20q. Proc Natl Acad Sci USA 1991;88:1484–1488.

SECTION II

The NON mouse as an animal model for Type II diabetes mellitus

CHAPTER 4

Diabetic Syndrome in the NON mouse

GOJI HASEGAWA[a], MASAYUKI HATA[a], KOJI NAKANO[a], MOTOHARU KONDO[a] and TAKAHIRO KANATSUNA[b]

[a]*First Department of Internal Medicine, Kyoto Prefectural University of Medicine, Kyoto, Japan*
[b]*Department of Metabolic Disease, Kyoto City Hospital, Kyoto, Japan*

Current concepts of a new animal model: The NON mouse
Edited by N. Sakamoto, N. Hotta and K. Uchida

Contents

Introduction

Non-insulin-dependent diabetes (NIDDM) is the more common form of diabetes mellitus, but its pathogenesis remains unclear and controversial. Many rodents with spontaneously hyperglycemic syndrome have been reported as models of human NIDDM, and they are classified into two groups according to whether they have obesity or not [1]. However, it must be remembered that no animal model of diabetes corresponds perfectly to human NIDDM. In this chapter, we report the diabetic syndrome and insulin secretion of the NON mouse, which has potential usefulness as a model for human NIDDM.

Materials and methods

Animals

Six-week-old male and female NON mice were provided from Aburahi Laboratories, Shionogi Co. Ltd., Shiga, Japan. Age matched ICR mice of both sexes were purchased from Charles River Japan (Kanagawa, Japan), and were used as controls. All mice were maintained on ordinary mouse chow (Oriental Yeast, Tokyo, Japan), and water ad libitum.

Experimental protocol

The body weight and fed blood glucose levels (0900 to 1100 h) were measured every two weeks. Urine glucose was also examined every two weeks by TES-TAPE® (ELI LILLY Co., Indianapolis, USA). Glucose tolerance test was performed at 10 and 16 weeks of age. Pancreas perfusion study was carried out at 16 weeks of age.

(a) Glucose tolerance test

The mice were fasted overnight and injected intraperitoneally with a 20% glucose solution at a dose of 2 mg/g·BW. Blood samples were collected without anesthesia at 0, 30, 60, and 120 minutes after the injection. Plasma insulin levels were measured at 0 and 30 minutes.

(b) Pancreas perfusion

The technique has been described previously [2, 3]. The basal perfusate was Krebs-Ringer bicarbonate buffer (pH = 7.4) with 1% bovine serum albumin, 3% Dextran T70 and 5 mM glucose, equilibrated with a continuous supply of 95% O_2/5% CO_2 gas. The pancreas was perfused at 37° C, at a flow rate of 0.6 ml/min. After 20 minutes of equilibration with the basal perfusate, the pancreas was stimulated with 30 mM glucose for 10 minutes. The pancreas was re-equilibrated with basal perfusate for 10 minutes, and then stimulated with 19 mM arginine in the presence of 5 mM glucose for 10 minutes. Samples were collected from the portal vein at 1 or 2 minute intervals. After the perfusion, the pancreas was homogenized in acid-alcohol solution for the measurement of the insulin concentration.

(c) Analytical methods

All blood samples were obtained from the orbital vein plexus and glucose levels were measured by the glucose oxidase method (FUJI DRI-CHEM SYSTEM, Fuji Medical System Co., Tokyo, Japan). Insulin levels in the perfusate, pancreatic extract, and plasma were assayed by solid phase radioimmunoassay (insulin RIA kit, Dinabot Radioisotope Lab., Tokyo, Japan).

Data presentation and statistical method

The insulin release from the perfused pancreas per minute was calculated by multiplying the insulin concentration of each sample by the flow rate and expressed as μU/min. The total insulin release was calculated by adding the insulin release per minute during the perfusate condition and expressed as μU/10 min.

The results are shown as the mean ± SEM. Statistical significance was determined using the Student's t-test.

Results

Body weight (Fig. 1)

Male NON mice were significantly lighter in body weight than ICR mice throughout the study. There was no difference between the two in female mice.

Fed blood glucose level (Fig. 2) and urine glucose

Mild hyperglycemia was observed in NON mice of both sexes, although the values fluctuated reaching normal levels (Male; NON 143.0 ± 5.4, ICR 125.8 ± 3.4 mg/dl $P < 0.02$: Female ; NON 159.3 ± 10.0, ICR 133.3 ± 8.1 mg/dl $P < 0.05$ at 16 weeks of age). Glucosuria was not detected in NON mice.

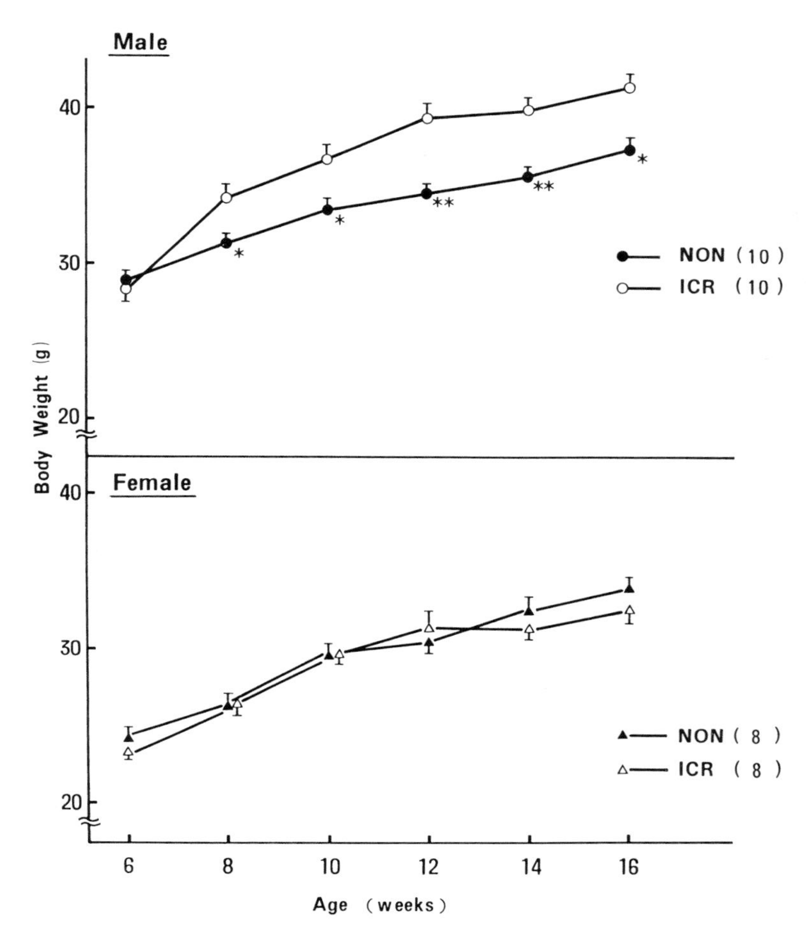

Fig. 1. Changes in body weight of NON and ICR mice. The values in parentheses were the number of animals tested. Values represent mean ± SEM. $^*P < 0.01$, $^{**}P < 0.001$.

Glucose tolerance test (Fig. 3)

Glucose tolerance was mildly impaired in male NON mice at 10 and 16 weeks of age, demonstrating significantly higher glucose levels than controls at 60 and 120 minutes after the glucose loading. In female NON mice, impaired glucose tolerance appeared only at 16 weeks of age. Plasma insulin responses after glucose loading were significantly reduced in NON mice of both sexes at 16 weeks of age ($P < 0.05$).

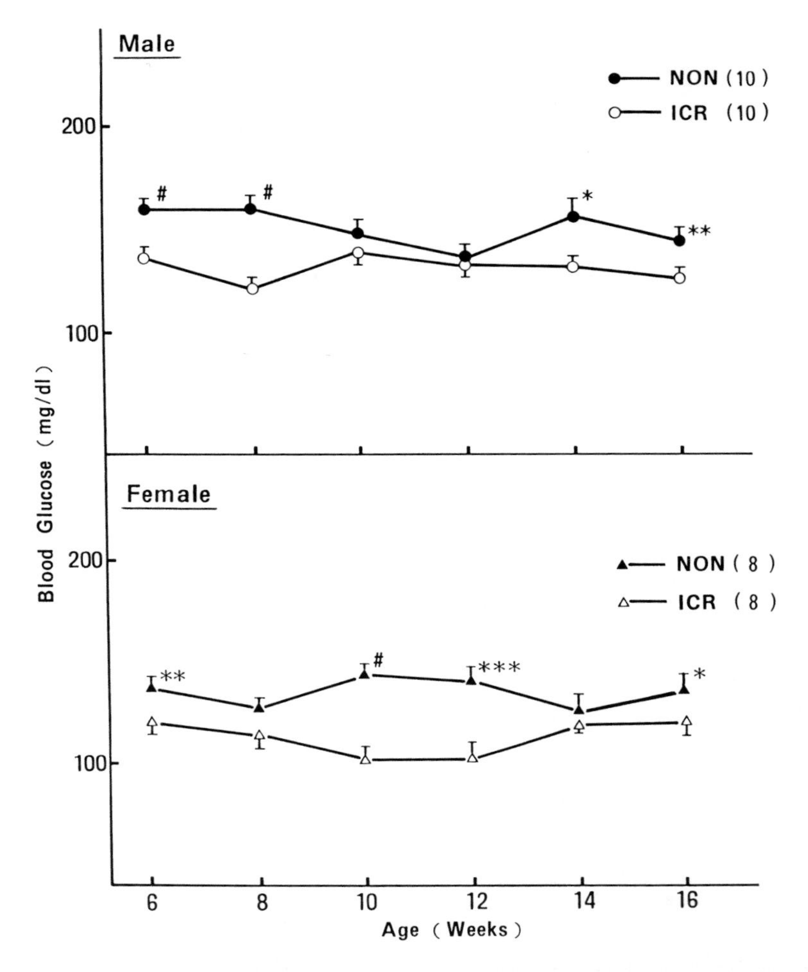

Fig. 2. Fed blood glucose levels in NON and ICR mice. The values in parentheses were the number of animals tested. Values represent mean ± SEM. $*P < 0.05$, $**P < 0.02$, $***P < 0.01$, $^{\#}P < 0.001$.

Pancreas perfusion and pancreatic insulin concentration (Fig. 4 and Table I)

In the pancreas perfusion studies at 16 weeks of age, there were no differences in the insulin release at 30 mM glucose between NON and ICR mice, regardless of sexes. The arginine induced insulin release at 5.0 mM glucose in NON mice was also similar to that of ICR mice.

There were no differences in pancreatic insulin concentration between NON and ICR mice.

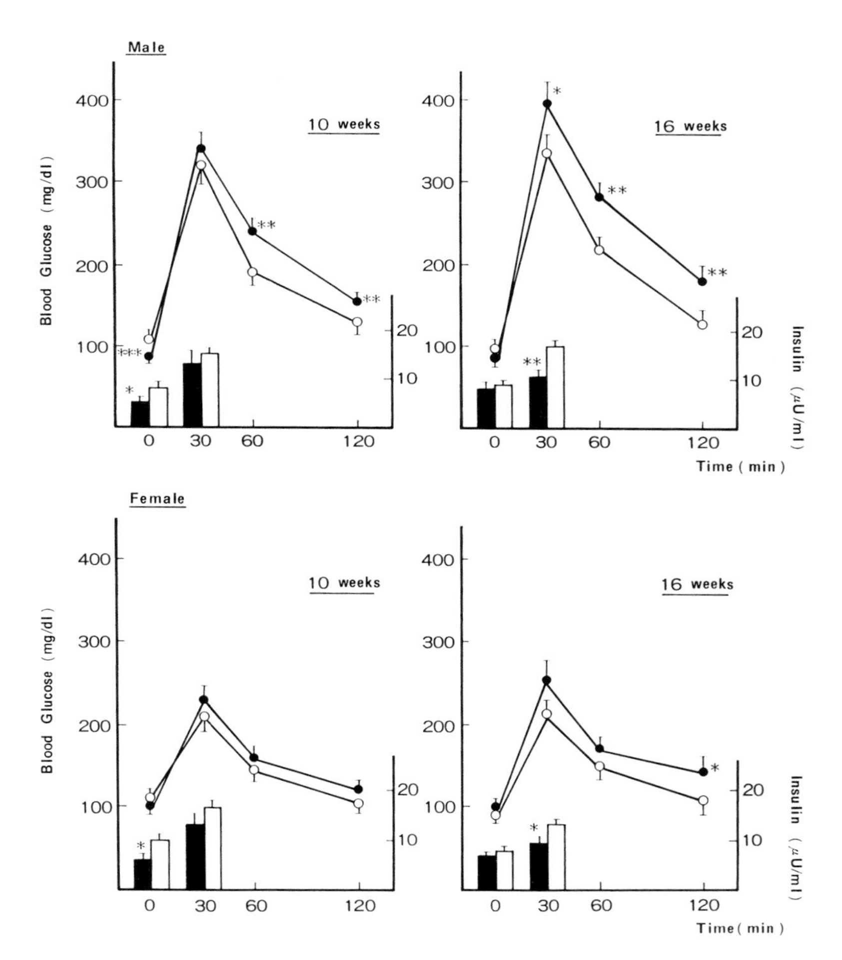

Fig. 3. Glucose tolerance and plasma insulin response after intraperitoneal injection of 2 mg/g·BW glucose in NON and ICR mice. The line graphs indicate the changes of blood glucose levels, and the bar graphs indicate the plasma insulin response. ●— NON (♂; $n = 8$; ♀; $n = 5$) ○— ICR (♂; $n = 8$, ♀; $n = 5$), ■ NON (♂; $n = 5$, ♀; $n = 5$) □ ICR (♂; $n = 5$, ♀; $n = 5$). Values represent mean ± SEM. *$P < 0.05$, **$P < 0.02$, ***$P < 0.001$.

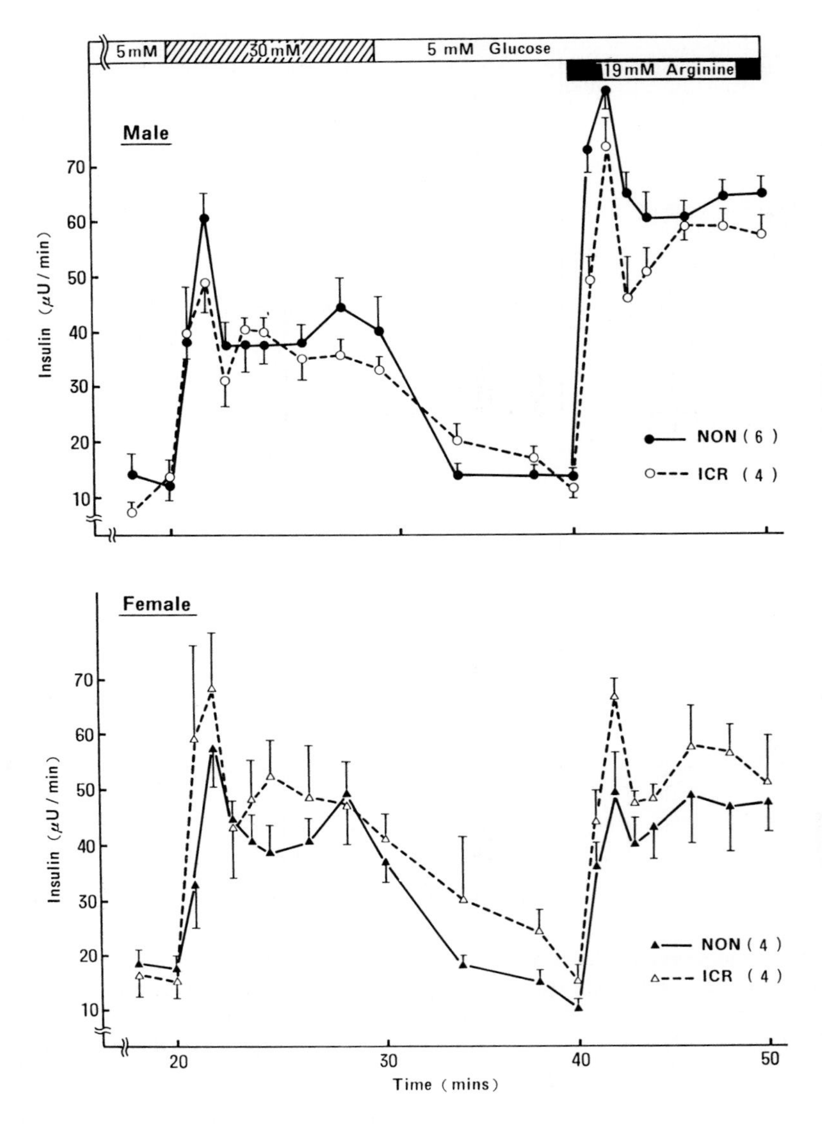

Fig. 4. Insulin response to 30 mM glucose and 19 mM arginine during pancreas perfusion. The pancreas perfusion was performed at 16 weeks of age. The values in parentheses are the number of animals tested. Values represent mean ± SEM.

TABLE I
Insulin release from the perfused pancreas and pancreatic insulin concentration in NON and ICR mice at 16 weeks of age. Values represent mean ± SEM

			Total insulin release (μU/10 mins)		Insulin concentration
	Mice	(*n*)	30 mM Glucose	19 mM Arginine	(U/g·pancreas)
♂	NON	(6)	333.2 ± 40.2	470.0 ± 30.5	0.77 ± 0.04
♂	ICR	(4)	303.7 ± 23.1	390.3 ± 30.7	0.85 ± 0.06
♀	NON	(4)	343.1 ± 50.1	310.7 ± 40.2	0.83 ± 0.05
♀	ICR	(4)	400.3 ± 90.2	371.6 ± 31.5	0.72 ± 0.05

Discussion

The diabetic syndrome of NON mouse is characterized as follows: (1) slightly, but significantly, elevated fed blood glucose levels from six weeks of age; (2) impaired glucose tolerance on ip-GTT with reduced insulin response; (3) NON mice did not show overt diabetes when they were maintained on ordinary mouse chow. The syndrome of this animal may be equivalent to impaired glucose tolerance (IGT) of humans.

On pancreas perfusion, NON mice did not show selective loss of insulin secretion to glucose which has been observed in human and animal diabetes [4, 5]. It has been suggested that chronic exposure of pancreatic β cells to high glucose levels could induce this defect [6, 7]. Thus, the blood glucose levels in NON mice may not be sufficient to induce the defect. However, special attention should be paid to the interpretation of the result of the present perfusion study. Because, the stimulation of 30 mM glucose may be too enormous to detect the impaired insulin secretory capacity of NON mice.

Reduced expression and restriction fragment length polymorphism of insulin gene [8] or abnormalities in insulin secretory function of β-cells have been reported in NON mice (may be described in other chapters). If some external factors, such as overeating or stress, could cause overt diabetes in NON mouse with these intrinsic β-cell abnormalities, this mouse will be more useful as a model of human NIDDM.

References

1 Mordes JP, Rossini AA. Animal models of diabetes mellitus. In: Marble A, Krall LP, Bradley RF, Christlieb AR, Soeldner JS, eds. Joslin's Diabetes Mellitus, Philadelphia: Lea & Febiger, 1985;110–137.

2 Hasegawa G, Mori H, Sawada M, Takagi S, Shigeta H, Kitagwa Y, Nakano K, Kanatsuna T, Kondo M. Overt diabetes induced by overeating in neonatally STZ-treated impaired glucose tolerant mice: Long term follow up study. Endocrinol Jpn 1989;36:471–479.

3 Hasegawa G, Mori H, Sawada M, Takagi S, Shigeta H, Kitagawa Y, Nakano K, Kanatsuna T, Kondo M. Dietary treatment ameliorates overt diabetes and decreased insulin secretion to glucose, induced by overeating in impaired glucose tolerant mice. Horm Meta Res 1990;22:408–412.

4 Unger RH, Grundy S. Hyperglycaemia as an inducer as well as a consequence of impaired islet cell function and insulin resistance: implications for the management of diabetes. Diabetologia 1985;28:119–121.

5 Giroix MH, Portha B, Kergoat M, Bailbe D, Picon L. Glucose insensitivity and amino-acid hypersensitivity of insulin release in rats with non-insulin-dependent diabetes: a study with the perfused pancreas. Diabetes 1983;32:445–451.

6 Kergoat M, Bailbe D, Portha B. Insulin treatment improves glucose-induced insulin release in rats with NIDDM induced by streptozocin. Diabetes 1987;36:971–977.

7 Leahy JL, Cooper HE, Deal DA, Weir GC. Chronic hyperglycemia is associated with impaired glucose influence on insulin secretion: A study in normal rats using chronic in vivo glucose infusions. J Clin Invest 1986;77:908–915.

8 Ohgaku S. Reduced expression and restriction fragment length polymorphism of insulin gene in NON mice: a new animal model for nonobese, noninsulin-dependent diabetes. In: Shafrir E, Renold AE, eds. Lessons from animal diabetes II. John Libbey, 1988: 319–323.

CHAPTER 5

Body fat accumulation and metabolic disturbance in the NON mouse

MASAYUKI HATA[a], GOJI HASEGAWA[a], NAOTO NAKAMURA[a], KOJI NAKANO[a], MOTOHARU KONDO[a] and TAKAHIRO KANATSUNA[b]

[a]First Department of Internal Medicine, Kyoto Prefectural University of Medicine, Kyoto, Japan,
[b]Department of Metabolic Disease, Kyoto City Hospital, Kyoto, Japan

Current concepts of a new animal model: The NON mouse
Edited by N. Sakamoto, N. Hotta and K. Uchida

Contents

Introduction

The NON mouse shows impaired glucose tolerance (IGT) without hyperinsulinemia and obesity, and its potential usefulness as a human non-obese NIDDM model has been discussed [1–3].

Little is known, however, about the pathogenesis of IGT in NON mice. To clarify this, we investigated the relationship between body fat accumulation and metabolic disturbance in the NON mouse.

Materials and methods

Animals

Four weeks of age (w.a.) male and female NON mice were provided from Aburabi Laboratories, Shionogi Co. Ltd., (Shiga, Japan). Age matched ICR mice of both sexes purchased from Charles River Japan (Kanagawa, Japan), and were used as controls. Both mice received ordinary mouse chow (Oriental Yeast, Tokyo, Japan) throughout the study. All mice were housed in a temperature-controlled room (20–23°C), with overhead lights, on from 0700–1900 h, and had free access to tap water.

Experimental protocol

The body weight and fed blood glucose levels (0900 to 1100 h) were measured every two weeks. Plasma triglyceride and cholesterol levels (0900 to 1100 h) were measured at 4, 10 and 20 w.a.

(a) Glucose tolerance test (ip-GTT)
Ip-GTT was performed at 4, 10 and 20 w.a. The mice were fasted overnight and injected intraperitoneally with a 20% glucose solution at a dose of 2 mg/g·BW. Blood samples were collected without anesthesia at 0, 30, 60, and 120 minutes after the injection. Plasma insulin levels were measured at 0 and 30 minutes.

(b) Lipoprotein lipase (LPL) and hepatic lipase (HL) activity post-heparin plasma (expressed as μmol[^{14}C]triolein/min/h)
LPL and HL activity were measured at 10 w.a. Mice were injected with heparin 10 Units/kg through the tail vein. Twenty minutes after the injection, the animals were decapitated and bled. Plasma was prepared and assayed for lipoprotein lipase and hepatic lipase [4–8]. LPL activity was determined using the methods of Nilsson-Ehle and Schotz [4]. The HL activity was determined using the same procedure in the presence of 2 M NaCl.

(c) Triglyceride secretion rate (TGSR) (expressed as mg/dl/min.)
TGSR was measured using Triton WR 1339 (Hanni Chem. Drug Co.) [9] at 10 w.a. of male mice.

(d) Body fat volume (expressed as mg/g·BW)
Body fat volume was measured using right inguinal subcutaneus (S), retroperitoneal (R), and mesentric (M) fat at 4, 10, and 20 w.a.

(e) Lee index
Lee index was measured at 10 and 20 w.a.

(f) Food consumption (expressed as Kcal/mouse/day)
Food consumption was measured three times each week, from 10 to 15 w.a.

(g) RMR
RMR was measured using closed-circuit metabolic system from 12 to 14 w.a.

(h) Plasma T3 levels
Plasma T3 levels were measured at 4, 10 and 20 w.a.

Analytical methods

All blood samples were obtained from the orbital vein plexus and plasma glucose levels were measured by the glucose oxidase method. Plasma triglyceride (T.G.) and cholesterol (Ch.) levels were measured by the LPL-GK-G3 PDH-D1 enzyme methods and the cholesterol oxidase methods (FUJI DRI-CHEM SYSTEM, Fuji Medical System Co., Tokyo, Japan), respectively.

Insulin levels in plasma were assayed by solid phase radioimmunoassay (insulin RIA kit, Dainabot Radioisotope Lab., Tokyo, Japan.) and T3 levels in plasma by solid phase radioimmunoassay (SPAC T3 RIA Kit, Daiichi radioisotope Lab., Tokyo, Japan).

Statistical analysis

The results were shown as the mean ± SEM. Statistical significance was determined using the Student's t-test.

Results

Metabolic disturbance

(a) Non-fasting blood glucose (Fig. 1)
Slightly elevated fed blood glucose levels were observed from 4 w.a. in NON mice of both sexes compaired to those in ICR mice.

(b) Glucose tolerance test (Fig. 2)
On ip-GTT, mild IGT was observed at 10 and 20 w.a. only in NON males. Further-

Non-fasting blood glucose

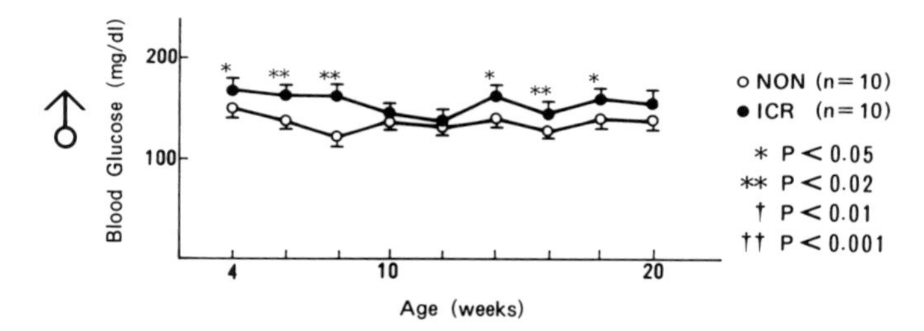

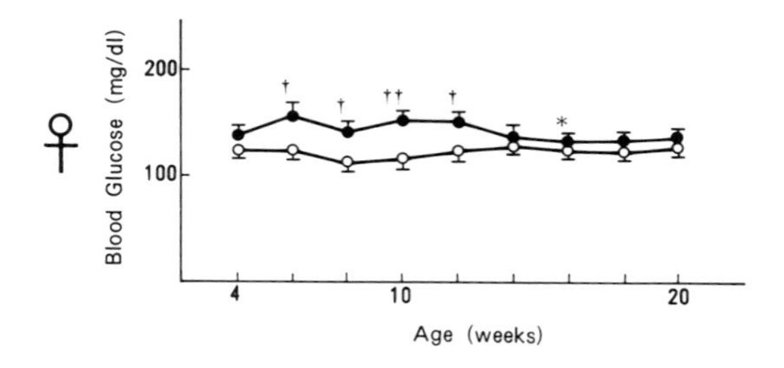

Fig. 1. Non-fasting blood glucose levels in NON and ICR mice of both sexes. Upper and lower panels indicate values in males and females, respectively. Number of animals in each group is given in parentheses. Values represent mean ± SEM.

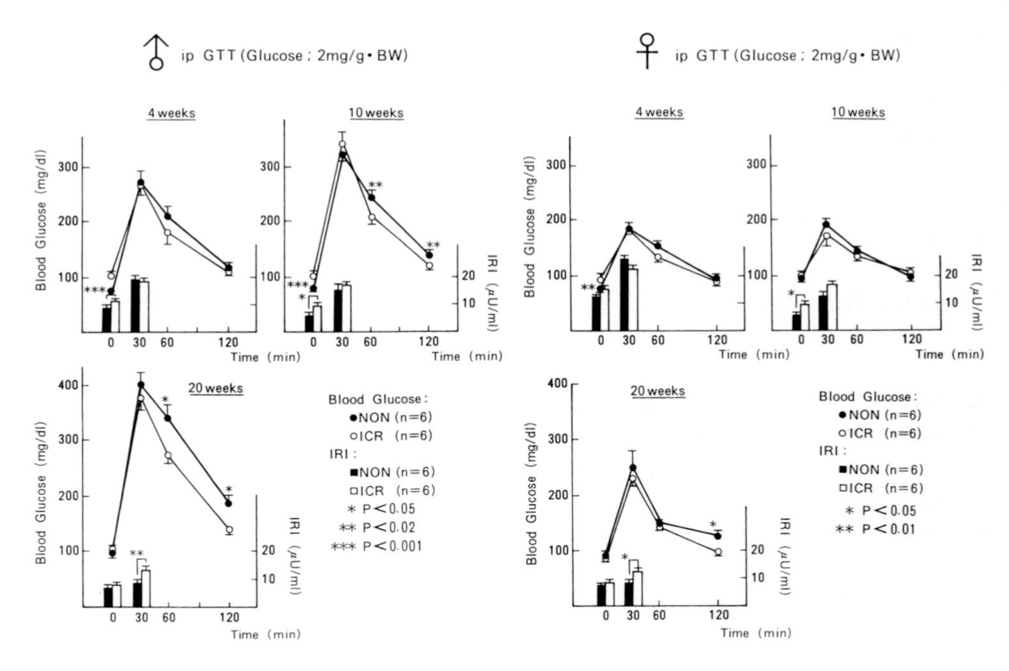

Fig. 2. Glucose tolerance and plasma insulin response after intraperitoneal injection of 2 mg/g·BW glucose in NON and ICR mice of both sexes at 4,10 and 20 w.a. The line graphs indicate the changes of blood glucose levels, and the bar graphs indicate the plasma insulin response. Left and right panels indicate values in males and females, respectively. Values represent mean ± SEM. Number of animals in each group is given in parentheses.

more, the fasting plasma insulin level was significantly lower at 4 and 10 w.a. in NON mice compared to that in ICR control mice and low insulin response was observed at 20 w.a.

(c) T.G. and Ch. (Fig. 3)

Hypertriglycedemia was observed from 4 w.a. in NON mice. On the other hand, plasma cholesterol levels of NON mice were significantly higher only at 4 w.a. than that of ICR mice.

LPL and HL activity in post-heparin plasma, and TGSR were not different between the NON and ICR groups.

Body fat volume (expressed as mg/g·BW) (Fig. 4)

The body fat volume including subcutaneus, retroperitoneal and mesentric fat significantly increased from 4 w.a. in NON males, nevertheless they were light in body

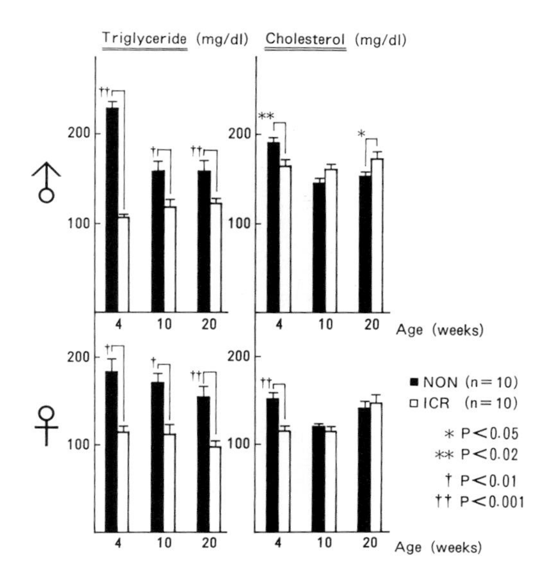

Fig. 3 Plasma triglyceride and cholesterol levels in NON and ICR mice of both sexes at 4,10 and 20 w.a. Upper and lower panels indicate values in males and females, respectively. Number of animals in each group is given in parentheses. Values represent mean ± SEM.

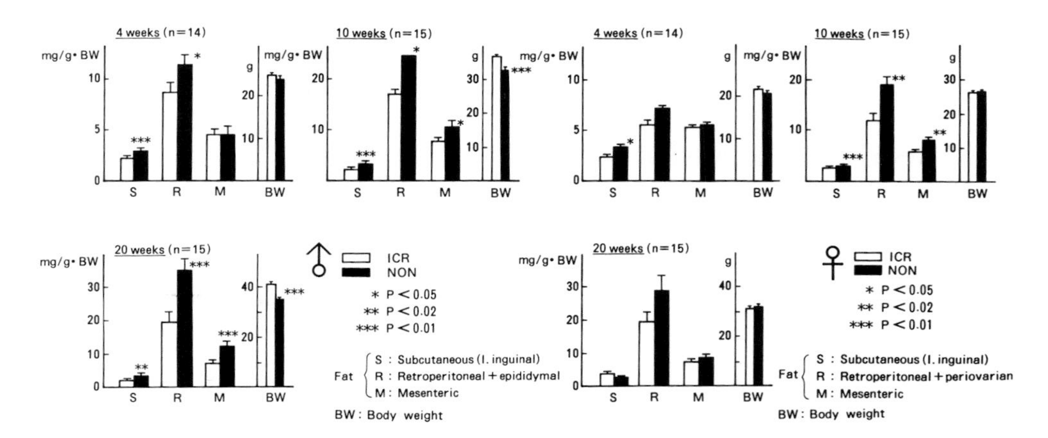

Fig. 4. Body fat volume (mg/g·BW) in NON and ICR mice of both sexes at 4,10 and 20 w.a. Left and right panels show bars in males and females, respectively. Number of animals in each group is given in parentheses. Values represent mean ± SEM.

weight. On the other hand, in NON females subcutaneous fat volume increased from 4 to 10 w.a., although the retroperitoneal and mesenteric fat volume significantly increased only at 10 w.a.

The Lee index (measured at 10 and 20 w.a.) in NON mice was similar to that in ICR mice.

Food consumption

There was no difference in food intake between NON and ICR mice.

RMR and plasma T3 levels

There were no differences in RMR and plasma T3 levels between NON and ICR mice.

Discussion

From this study, it is suggested that the NON mouse is characterized by mild IGT, hypertriglycedemia and increase in body fat volume regardless of its normal or light body weight.

The relationship between visceral fat accumulation and metabolic disorders has become of major interest recently [10–11]. So the NON mouse might become a useful model for these disorders. In humans, generally, hyperinsulinemia induced by overeating causes the increase of body fat accumulation, whereas in the NON mouse the increase of body fat accumulation is induced in spite of low insulin-response to glucose-loading. It is also well known that insufficient insulin-action causes the lack of LPL activity, and consequently hyperchylomicronemia and finally hypertriglycedemia [12]. On the other hand, NIDDM patients with obesity and hyperinsulinemia can preserve LPL activity, and have hypertriglycedemia by over-production of VLDL [12]. The NON mouse, however, had almost normal LPL activity and TGSR regardless of hypertriglycedemia. And there were no significant differences in the food intake, plasma T3 levels and RMR between NON and ICR mice. Therefore, lipoprotein contents, food absorption, other hormonal actions and genetic factors should be investigated to clarify the mechanism of the body fat accumulation and hypertriglycedemia.

The NON mouse has a unique character of the increase in body fat volume without obesity and hyperinsulinemia, which is different from the conventional obese-diabetic animal models [13–17].

References

1 Ogaku S. et al. NON mice: a new animal model for non-obese NIDDM, Diabetes Res Clin Pract 1985;6(1):415.
2 Komeda K et al. Mandible analysis of NOD and NON strains of mice. Lab Anim 1984;18:237–242.
3 Makino S, et al. Breeding of a non-obese, diabetic strain of mice, Exp Anim 1980;29(1), 1–13.
4 Nilsson-Ehle P, Schotz M.C. A stable, radioactive substrate emulsion for assay of lipoprotein lipase. J Lipid Res 1976;17:536–541.
5 Bucolo G, David H. Quantitative determination of serum triglycerides by the use of enzymes. Clin Chem 1973;19:476–482.
6 Kozlowska A, Sadurska B, Szymczyk T. Effect of dichlorvos on the activity of lipoprotein lipase from adipose tissue, on plasma lipids and post-heparin lipolytic plasma activity in rats. Arch Toxicol 1988;62:227–229.
7 Pykälistö O, Vogel WC, Bierman EL. The tissue distribution of triglycerol lipase, monoacylglycerol lipase and phospholipase A in fed and fasted rats. Biochem Biophys Acta Amsterdam: Elsevier 1974;369:254–263.
8 Williams SP, Johnson EA. Release of lipoprotein lipase and hepatic triglyceride lipase in rats by heparin and other sulphated polysaccharides. Thromb Res 1989;55:361–368.
9 Robertoson RP, Gavarski DJ, Henderson JD, Porte D Jr., Bierman EL. Accelerated triglyceride secretion. A metabolic consequence of obesity. J Clin Invest 1973:52:1620–1626.
10 Kissebah AH, Vydelingum N, Murray R, et al. Relation of Body fat distribution to metabolic complications of obesity. J Clin Endocrinol Metab 1982;54:254.
11 Fujioka S, Matsuzawa Y, Tokunaga K, et al. Contribution of intra-abdominal fat accumulation to the impairement of glucose and lipid metabolism in human obesity. Metabolism 1987;36:54.
12 Abrums JJ, Ginsberg H, Grumdy SM. Metabolism of cholesterol and plasma triglycerides in non-ketotic diabetes mellitus. Diabetes 1982;31:903.
13 Tachino Y. Spontaneously diabetic non obese (NOD) mouse with insulitis. Anim Diabetes 1990;4:19–38.
14 Ikeda H, Shino A, Matsuo T, et al. A new genetically obese-hyperglycemic rat (Wister Fatty). Diabetes 1981;30:1045–1050.
15 Sugiyama Y, Shimura Y, Ikeda H. Pathogenesis of hyperglycemia in genetically obese-hyperglycemic rats, Wister Fatty: Presence of hepatic insulin resistance. Endocrinol Jpn 1989;36(1):65–73.
16 Bray GA, York DA. Genetically transmitted obesity in rodents. Physiol Rev 1971;51(3):598–646.
17 Sugiyama Y, Shimura Y, Ikeda H. Effects of pioglitazone on hepatic and peripheral insulin resistance in Wister Fatty Rats. Arzneim-Forsch:/Drug Res 1990;400:436–440.

CHAPTER 6

Cytoplasmic Ca^{2+} response in pancreatic β cells of the NON mouse

KAZUO TSUJI[a], TOMOHIKO TAMINATO[b], HITOSHI ISHIDA[c], YOSHIMASA OKAMOTO[a], YOSHIYUKI TSUURA[a], SEIKA KATO[c], TAKESHI KUROSE[c], YUTAKA SEINO[c] and HIROO IMURA[a]

[a]*Second Division, Department of Internal Medicine, Kyoto University School of Medicine, Kyoto, Japan*
[b]*Second Department of Medicine, Hamamatsu University School of Medicine, Hamamatsu, Japan*
[c]*Department of Metabolism and Clinical Nutrition, Kyoto University School of Medicine, Kyoto, Japan*

Current concepts of a new animal model: The NON mouse
Edited by N. Sakamoto, N. Hotta and K. Uchida

Contents

Introduction

In non-insulin-dependent diabetes mellitus (NIDDM), insulin secretion has been demonstrated to be selectively impaired to glucose [1]. Leahy et al. [2] have reported that impairment of glucose sensitivity in vivo evolves in normal rats after chronic glucose infusions. Furthermore, NON mice have been originally derived from the CTS subline of the ICR strain with cataract and small eyes, and have also been speculated to be an appropriate model of NIDDM [3]. Using the isolated perfused pancreas, Kano [4] has reported that the first phase insulin secretion in response to 30 mM glucose was significantly decreased in NON mice compared to control ICR mice. It is widely accepted that cytoplasmic free calcium ($[Ca^{2+}]_i$) plays an important role in insulin secretion from the pancreatic β cells [5]. The mechanism of modulating $[Ca^{2+}]_i$ by glucose in the β cells involves depolarization of the β cells by closing the ATP-sensitive K^+ channels followed by entry of free calcium through voltage-dependent calcium channels (VDCC) [6]. On the other hand, carbachol has been demonstrated to possess a different manner of elevating $[Ca^{2+}]_i$ and to stimulate inositol trisphosphate (1,4,5-IP_3)-induced Ca^{2+} release from the intracellular calcium pool [7]. However, the relationship between the intracellular Ca^{2+} handling and insulin secretion from the β cells of NIDDM has not been elucidated yet. In order to characterize the effects of glucose and carbachol on $[Ca^{2+}]_i$ dynamics as well as insulin release from the pancreatic β cells, we evaluate their interrelationship in NON mice in the present study.

Materials and methods

The NON mice were obtained from the Aburahi Laboratories, Shionogi and Co. (Shiga, Japan). They were fed ad libitum and had free access to water until 10 weeks old. Then, these animals were divided into two groups. One group was administered orally 10% glucose dissolved in water for 8 weeks in order to deteriorate their glucose tolerance (G group), and the other was administered water alone (W group). At 18 weeks old, NON mice of both groups were used for the following experiments.

Intraperitoneal glucose tolerance test (IPGTT)

After the overnight fast, the mice were given 2 mg/g body weight glucose intraperitoneally. The blood samples were obtained by snipping the tail vein at 0, 15, 30, 45, 60, 90 and 120 minutes after the gluocose loading to determine blood glucose levels.

Insulin release study

Pancreatic islets were isolated from the animals of both groups in fed state by the collagenase digestion method according to Goto et al. [8]. Groups of five freshly isolated islets were preincubated at 37°C for 30 min. in Krebs-Ringer bicarbonate buffer (KRBB) saturated with 95% O_2–5% CO_2 (pH 7.4), containing 3.3 mM glucose and 0.2% bovine serum albumin (preincubation medium). These groups of islets were then incubated at 37°C for 30 min. in the media which contains each of insulin secretagogues (incubation medium), as listed in Table I. The released insulin was measured by radioimmunoassay using rat insulin (Novo, Bagsvaerd, Denmark) as standard. The bound and free hormones were separated by the polyethylene glycol method [9].

Measurement of cytoplasmic concentration of free calcium (Ca^{2+})

Freshly isolated islets were dispersed by gentle pipetting to small clusters of islet cells which contained about 10–50 cells in each. The clusters of islet cells were washed two times by centrifugation in RPMI 1640 medium containing 3.3 mM glucose. The viabilities of dispersed islet cells of both groups were above 95% by the trypan blue method. Then, fifty microliters of the cell suspension was plated onto glass coverslips which had been coated with Cell-TaK® (Collaborative Research Inc., MA, USA), and incubated for 30–70 min. to allow the clusters to adhere on the slips. The cells were loaded for 30 min, at 37°C in KRBB with 1 μM (final concentration) of fura-2-acetoxymethylester (Molecular Probes, Eugene, OR, USA) dissolved in Me_2SO_4.

TABLE I
Insulin release from pancreatic islets of male NON mice. Values are means ± S.E.

Stimuli	W group	G group	*P*
Glucose 3.3 mM	10.3±2.4	3.5±0.9	<0.05
Glucose 8.3 mM	16.7±2.1	6.4±1.0	<0.01
Glucose 16.7 mM	15.1±3.5	5.8±1.6	<0.05
Arginine 20 mM (Glucose 8.3 mM)	33.7±2.5	25.9±4.2	N.S.

After washing fura-2 with preincubation medium the coverslips with the loaded cells were placed on the stage of inverted microscope and covered with coverglass. The superfusion of the attached islet cells were performed by manual pipetting, and the superfused medium was suctioned from another side of the coverslip by sufficient filter papers. Each of various secretagogues was superfused at 37°C by heating the medium and the stage of microscope. Dual excitation wavelength measurement (340 and 360 nm) was performed by an automatic device allowing two wavelengths to be alternatively selected using microcomputer-driven shutters. The clusters of islet cells were viewed with a ×20 fluorescence objective (Nikon, Tokyo, Japan). Fluorescence emission at 510 nm was monitored with a silicon-intensified target camera (C2400-08H, Hamamatsu Photonics, Hamamatsu, Japan), and the 340/360 ratio calculation was digitalized every 10 s by a computerized image processor (ARGUS 100/CA, Hamamatsu Photonics). The corresponding ratio image was pseudo-colored according to $[Ca^{2+}]_i$ levels. Since it has been observed that the insulin response to glucose is remarkably reduced when the islet cells are completely dispersed [10], here we used mildly but not completely dispersed islet cells for the determination of $[Ca^{2+}]_i$ in order to investigate the effect of glucose on the intracellular events in a signal transduction system. In addition, it has been found, as previously reported [11], that the non-insulin containing islet cells of rats were characterized by their smaller size (< 10 μm in diameter). The experiments were always, therefore, performed on individual cells > 10 μm and kept in contact with other cells. The 340/360 nm fluorescence ratio of square area which covers approximately 2 to 3 β cells were used for calculating $[Ca^{2+}]_i$ with a standard procedure described above.

The lag times to various stimuli which evoke decrease (Lag D) or increase (Lag I) and the values of decrease or increase of the $[Ca^{2+}]_i$ of β cell clusters were measured, as shown in Fig. 1. The ratio of responsive cells which showed changes of $[Ca^{2+}]_i$ were calculated as a percentage.

Results are expressed as the mean value (± S.E.). The statistical differences were assessed by using the non-paired Students' t-test.

Results

Intraperitoneal glucose tolerance test (Fig. 2)

In male NON mice, the G group exhibited the deterioration of glucose tolerance in intraperitoneal glucose tolerance test. Fourty-five minutes after the glucose loading, blood glucose levels of the G group were found to be significantly higher compared to the W group ($P<0.05$). At 120 min. after, blood glucose of the G group was 287 ± 48 (mg/dl), being significantly higher than the value of 94 ± 2 in the W group ($P<0.01$). On the other hand, since the G group of female NON mice showed no

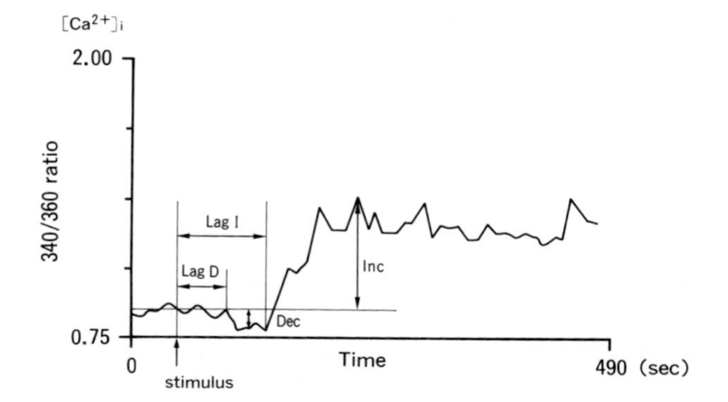

Fig. 1. Four parameters for statistical analysis of intracellular calcium concentration. After 60 s of equilibration period, 16.7 mM glucose or 100 μM carbachol was introduced to β-cell clusters. As indicated, Lag D and Lag I were the lag times to initiate decrease or increase of intracellular calcium, respectively. Dec and Inc were the values of the decrease or increase of intracellular calcium, respectively.

glucose intolerance when compared to the corresponding W group (data not shown), we examined only male NON mice in further experiments.

Insulin release study

The insulin release in the presence of various glucose concentrations from the isolated

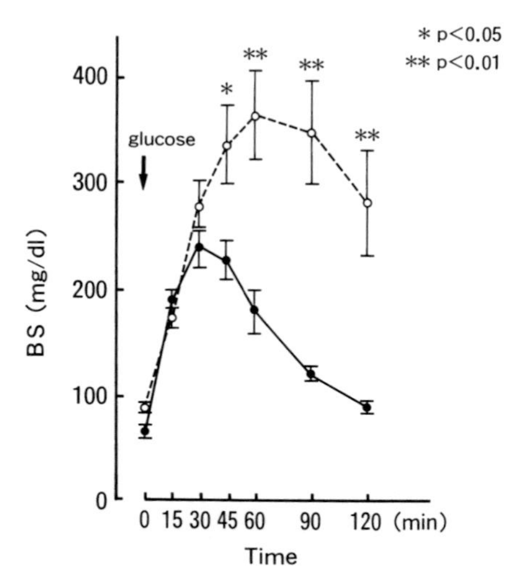

Fig. 2. Intraperitoneal glucose tolerance (2 mg/g body weight) of male NON mice. ●——●: NON mice given water alone (W group). ○---○: NON mice given 10% glucose dissolved in water (G group).

islets of the G group were diminished compared to islets obtained from the W group ($P<0.05$). In contrast, the insulin response to 20 mM arginine in the presence of 8.3 mM glucose was not statistically different between two groups (Table I).

Measurement of cytoplasmic Ca^{2+} concentration

Glucose-induced responses of $[Ca^{2+}]_i$ in β-cell clusters of both groups were variable. A considerable proportion of cells showed initially a slight lowering of $[Ca^{2+}]_i$ when the glucose concentration was raised from 3.3 to 16.7 mM followed by a gradual and large increment of $[Ca^{2+}]_i$. Representative $[Ca^{2+}]_i$ responses to 16.7 mM glucose in rats of both groups are shown in Fig. 3. Other cells show monophasic response, either decrease or increase of $[Ca^{2+}]_i$. More than half of the cells showed no response to glucose stimulation in both groups. After 16.7 mM glucose stimulation, the initial decrease of $[Ca^{2+}]_i$ in the G group was significantly larger than that of the W group ($P<0.05$). In contrast, the increase of $[Ca^{2+}]_i$ in response to glucose was not statistically different between two groups (Fig. 4). Both Lag D and Lag I in the G group were significantly longer than those in the control W group (Lag D: 47 ± 12 (s) versus 13 ± 9; $P<0.05$, and Lag I: 125 ± 11 versus 81 ± 10; $P<0.01$). The percentage of cells which showed initial lowering of $[Ca^{2+}]_i$ after glucose stimulation was higher in the G group compared to the W group (52% versus 27%). The percentage of cells showing increase of $[Ca^{2+}]_i$ were also higher in the G group (96% versus 69%).

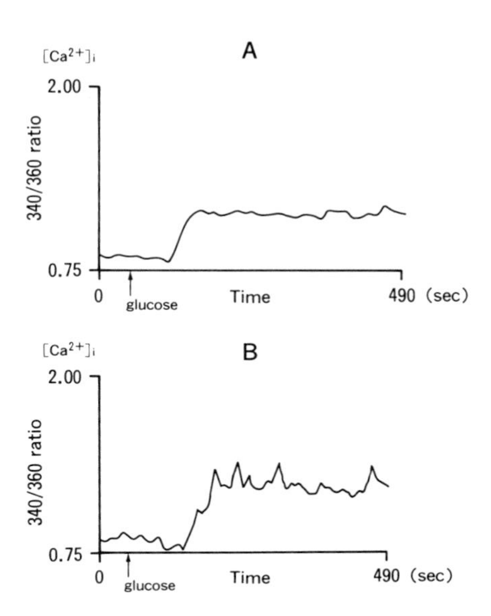

Fig. 3. The representative intracellular calcium responses to 16.7 mM glucose in β cells from the W group (A) and the G group (B).

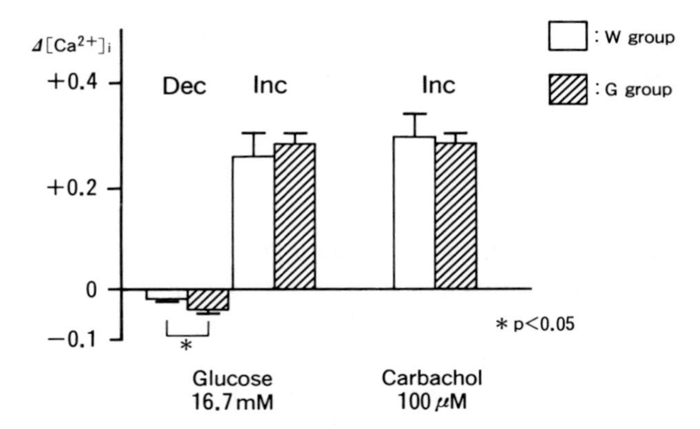

Fig. 4. The comparison of intracellular calcium responses to 16.7 mM glucose and 100 μM carbachol between the W and G groups. The decrease (Dec) of intracellular calcium in response to 16.7 mM glucose was significantly enhanced in the G group ($P<0.05$). On the other hand, the increase (Inc) was not statistically different in response to 16.7 mM glucose and 100 μM carbachol between the two groups.

One hundred μM carbachol in the presence of 3.3 mM glucose elicited a prompt and large increase of $[Ca^{2+}]_i$ in all the cell clusters examined. The representative case is shown in Fig. 5. The degree of increase was nearly equal in response to carbachol in both groups (Fig. 4). The lag time for the increase was slightly delayed in the G

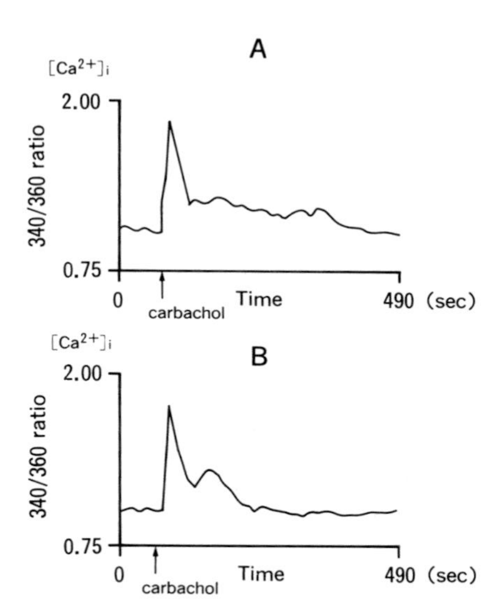

Fig. 5. The representative case of intracellular calcium responses to 100 μM carbachol in the W group (A) and the G group (B).

group compared to the W group, but not statistically different. In carbachol stimulation, nearly all cells showed prompt $[Ca^{2+}]_i$ elevation in both groups.

Discussion

The present study has shown the selective impairment of the glucose-induced insulin release from the isolated islets of the glucose-administered NON mice, as has been found in the neonatally streptozotocin-induced NIDDM rats [12]. On the other hand, the insulin release of NON mice to arginine stimulation was not impaired. Concerning the $[Ca^{2+}]_i$ response to glucose stimulation, the initial decrease of $[Ca^{2+}]_i$ of the glucose-administered NON mice was larger than that of mice given water alone (controls) but the subsequent increase of $[Ca^{2+}]_i$ was almost equal in β cells of both groups. In addition, the glucose-administered NON mice showed the significant delay in lag times to decrease and increase of $[Ca^{2+}]_i$ in response to glucose compared to controls. On the contrary, the increases of $[Ca^{2+}]_i$ to carbachol were not observed to be impaired and the lag times of both groups were not statistically different. It is strongly suggested, therefore, that the selective impairment of glucose stimulated insulin release in the glucose-administered NON mice is at least in part due to the prolonged lag time and large initial decrease of $[Ca^{2+}]_i$ selectively observed in the β cells after the glucose loading.

The intracellular mechanisms responsible for the delayed and enhanced initial decrease of $[Ca^{2+}]_i$ response in the glucose-administered NON mice is still unclear at present, but a brief increase in ATP sensitive K^+ channel activity has been reported immediately after the glucose stimulation [6, 13]. This phenomenon has been thought to induce β-cell hyperpolarization, which in turn lowers $[Ca^{2+}]_i$ [14, 15]. This initial decrease of $[Ca^{2+}]_i$ in β cells by glucose stimulation has been reported to possess the relationship with the initial inhibition of insulin release from the perifused rat islets [16]. The delayed and enhanced initial decrease of $[Ca^{2+}]_i$ might therefore be, the cause of the impaired insulin secretion from islets of the glucose-administered NON mice. The significance of $[Ca^{2+}]_i$ decrease in impaired insulin secretion in this animal model should be further clarified. The machinery of $[Ca^{2+}]_i$ elevation and insulin release to glucose stimulation has been recognized to be mediated mainly by supposed processes through glucose uptake and metabolism in β cells [17]. ATP is produced in the metabolism of glucose, and inhibits the ATP-sensitive K^+ channel. By the closure of this channel, β cells are thought to be depolarized and VDCC opens to increase $[Ca^{2+}]_i$ [6]. Therefore, it is also possible that there may be some disturbances in these processes for $[Ca^{2+}]_i$ elevation as the cause of delayed $[Ca^{2+}]_i$ increase. In pancreatic islets of NIDDM rat model, Portha et al. [18] have reported that the relative unresponsiveness to glucose is associated with a deficient islet glucose metabolism and that this defect is not due to gross alterations in the glycolytic pathway, but probably reflects the alteration in islet mitochondrial function. Giroix et al. [19] have further speculated the alteration in interdependency of $[Ca^{2+}]_i$ handling and mitochondrial oxidative events as a major determinant of the selective impairment of insulin secretion to glucose stimulation. The prolonged delay to initiate the decrease and increase of $[Ca^{2+}]_i$ observed in the glucose-administered NON mice might be due to the delayed glucose metabolism. Because the increase of $[Ca^{2+}]_i$ of the glucose-administered NON mice to glucose stimulation was nearly equal to controls, the derangement in the processes distal to $[Ca^{2+}]_i$ elevation, for example, the reduced sensitivity to $[Ca^{2+}]_i$ of the supposed calcium-activated effector systems that control the insulin secretory process [20], might be another possibility for the mechanisms of impaired insulin release.

The increase of $[Ca^{2+}]_i$ to 100 μM carbachol was shown to be nearly identical in the two groups. This evidence might indicate that the system of Ca^{2+} release from 1,4,5-IP_3-sensitive intracellular Ca^{2+} pool is not disturbed in this model.

In conclusion, the most characteristic features of $[Ca^{2+}]_i$ response to glucose stimulation in islet β cells of the glucose-administered NON mice are the enhancement of initial decrease and the sluggish response of $[Ca^{2+}]_i$. This evidence might be closely related to the selective impairment of glucose stimulated insulin release from pancreatic β cells observed in this NIDDM model.

Summary

NON mice have been found and established from ICR mice with cataract and small eyes, and also been speculated to be one of the appropriate models of non-insulin-dependent diabetes mellitus (NIDDM). In this study, we investigated the characteristics of insulin secretion and dynamics of intracellular calcium concentration in pancreatic β cells of this model. NON mice aged 10 weeks were divided ino two groups. To deteriorate their glucose tolerance, 10% glucose was administered orally in one group for 8 weeks (G group), while the other was given water alone (W group). Male mice, but not female, exhibited the decreased glucose tolerance in intraperitoneal glucose tolerance test (IPGTT: 2 mg/g body weight). At 120 min. after the glucose load, plasma glucose was 287 $\pm$ 48 (M $\pm$ SE) (mg/dl) in the G group, significantly higher than the value of 94 $\pm$ 2 in the W group ($P<0.01$). Insulin secretion from the isolated mouse islets in the G group was significantly impaired in response to glucose of various concentrations examined (3.3 mM, 8.3 mM, 16.7 mM; $P<0.05$, $P<0.01$ and $P<0.05$, respectively). On the other hand, insulin response to 20 mM arginine in the presence of 8.3 mM glucose was not significantly different between two groups. After glucose stimulation, the intracellular calcium was transiently decreased, and then gradually increased in most of β cell clusters of both groups. The lag times for both decrease and increase in intracellular calcium were significantly longer in the G group compared to the W group ($P<0.01$). In addition, the degree of decrease was also significantly marked in the G group ($P<0.05$). On the other hand, the increase was not statistically different in response to both of 16.7 mM glucose and 100 μM carbachol between two groups. The insulin response was observed to be selectively decreased to glucose stimulation after long-term loading of glucose in this model. The large decrease and sluggish increase of intracellular calcium of β cells might be the cause of the selective impairment of insulin release in response to glucose stimulation in NON mice.

Acknowledgements

The authors thank Ms. Yoko Yoshihara and Ms. Hiroko Tachikawa for their careful secretarial work, and Mr. Yasuki Nomura for his technical assistance. This study was supported by Grants-in-Aid for Scientific Research 03671145 from the Ministry of Education, Science, and Culture, and from the Research Committee of Experimental Models for Intractable Diseases of the Ministry of Health and Welfare of Japan, and also by the grant provided by the Ichiro Kanehara Foundation.

References

1 Leahy JL. Natural history of β-cell dysfunction in NIDDM. Diabetes Care 1990;13:992–1010.
2 Leahy JL, Cooper HE, Weir GC. Impaired insulin secretion associated with near normoglycemia; study in normal rats with 96-h in vivo glucose infusions. Diabetes 1987;36:459–464.
3 Makino S, Kunimoto K, Muraoka Y, Mizushima Y, Katagiri K, Tochino Y. Breeding of a non-obese, diabetic strain of mice. Exp Anim 1980;29:1–13.
4 Kano Y. Insulin secretion in NOD (non-obese diabetic) and NON (non-obese non-diabetic) mouse. J Kyoto Pref Univ Med 1988;97:295–308.
5 Prentki M, Matschinsky FM. Ca^{2+}, cAMP and phospholipid-derived messengers in coupling mechanisms of insulin secretion. Physiol Rev 1987;67:1185–1248.
6 Arkhammer P, Nilsson T, Rorsman P, Berggren PO. Inhibition of ATP-regulated K^+ channels precedes depolarization-induced increase in cytoplasmic free Ca^{2+} concentration in pancreatic β-cells. J Biol Chem 1987;262:5448–5454.
7 Berridge MJ, Irvine RF. Inositol trisphosphate, a novel second messenger in cellular signal transduction. Nature 1984;312:315–321.
8 Goto M, Maki T, Kiyoizumi T, Satomi S, Monaco AP. An improved method for isolation of mouse pancreatic islets. Transplantation 1985;40:437–438.
9 Desbuquois B, Aurbach GD. Use of polyethylene glycol to separate free and antibody-bound peptide hormones in radioimmunoassays. J Clin Endocrinol 1971;33:732–738.
10 Halban PA, Wollheim CB, Blondel B, Meda P, Niesor EN, Mintz DH. The possible importance of contact between pancreatic islet cells for the control of insulin release. Endocrinology 1982;111:86–94.
11 Pipeleers DG, in't Veld PA, Van De Winkel M, Maes E, Schuit FC, Gepts W. A new in vitro model for the study of pancreatic A and B cells. Endocrinology 1985;117:806–816.

12 Tsuji K, Taminato T, Usami M, Ishida H, Kitano N, Fukumoto H, Koh G, Kurose T, Yamada Y, Yano H, Seino Y, Imura H. Characteristic features of insulin secretion in the streptozotocin-induced NIDDM rat model. Metabolism 1988;37:1040–1044.
13 Trube G, Rorsman P, Ohno-Shosaku T. Opposite effects of tolbutamide and diazoxide on the ATP-dependent K^+ channel in mouse pancreatic β-cells. Pflügers Arch 1986;407:493–499.
14 Rorsman P, Abrahamsson H, Gylfe E, Hellman B. Dual effects of glucose on the cytosolic Ca^{2+} activity of mouse pancreatic β-cells. FEBS Lett 1984;170:196–200.
15 Arkhammar P, Nilsson T, Berggren PO. Glucose-induced changes in cytoplasmic free Ca^{2+} concentration and the significance for the regulation of insulin release; measurements with fura-2 in suspensions and single aggregates of mouse pancreatic β-cells. Cell Calcium 1989;10:17–27.
16 Hellman B, Berne C, Grapengiesser E, Grill V, Gylfe E, Lund PE. The cytoplasmic Ca^{2+} response to glucose as an indicator of impairment of the pancreatic β-cell function. Eur J Clin Invest 1990;20:S10–S17.
17 Malaisse-Lagae F, Mathias PCF, Malaisse WJ. Gating and blocking of calcium channels by dihydropyridines in the pancreatic B-cell. Biochem Biophys Res Commun 1984;123:1062–1068.
18 Portha B, Giroix MH, Serradas P, Welsh N, Hellerström C, Sener A, Malaisse WJ. Insulin production and glucose metabolism in isolated pancreatic islets of rats with NIDDM. Diabetes 1988;37:1226–1233.
19 Giroix MH, Sener A, Bailbe D, Portha B, Malaisse WJ. Impairment of the mitochondrial oxidative response to D-glucose in pancreatic islets from adult rats injected with streptozotocin during the neonatal period. Diabetologia 1990;33:654–660.
20 Watkins DT. Ca^{2+}-calmodulin-dependent phosphorylation of islet secretory granule proteins. Diabetes 1991;40:1063–1068.

CHAPTER 7

Changes of the K^+ channel of the pancreatic B cell in the NON mouse

T. TAMINATO[a], T. YAMAZAKI[a], T. TOMINAGA[a], T. YOSHIMI[a], M. NISHIMURA[b], K. TSUJI[c] and Y. SEINO[c]

[a]*The Second Department of Medicine, Hamamatsu University School of Medicine, Hamamatsu, Japan*
[b]*Institute for Experimental Animals, Hamamatsu University School of Medicine, Hamamatsu, Japan*
[c]*The Second Department of Medicine, Kyoto University School of Medicine, Kyoto, Japan*

Current concepts of a new animal model: The NON mouse
Edited by N. Sakamoto, N. Hotta and K. Uchida

Contents

Introduction

The NON mouse, a subclone of the NOD mouse, was separated in Aburahi Laboratories, Shiga, Japan [1]. The animal was found to exhibit impaired glucose tolerance, mild hyperglycemia without obesity, whereas NOD mice are characterized by severe hyperglycemia, ketosis-proneness and insulitis. The latter has been thought to be a good model for insulin-dependent diabetes mellitus (IDDM), and the former for non-insulin dependent diabetes mellitus (NIDDM).

NIDDM, especially in early stage, is characterized by a selective loss of insulin secretion in response to glucose, while insulin secretion to amino acids or glucagon is preserved. Although the precise mechanism of the pancreatic B cell dysfunction in NIDDM remained unsolved, intrinsic defects such as an insensitivity to recognize glucose, coupling defectes between glucose recognition and secreting machinery and structural change in the secreting system have been postulated [2].

Pancreatic B cells respond to a high concentration of D-glucose with a consecutive electrical phenomenon, a slow initial depolarization and a rapid depolarization with action potential, caused by K^+ channel acivation and subsequent K^+ influx. Studies using $^{86}Rb^+$ as a tracer of K^+ demonstrated that exposure of glucose to the islets cause a reduction of $^{86}Rb^+$ efflux which may result in the initial depolarization [3]. Depolarization may be followed by voltage-dependent activation of the Ca^{2+} channel, a rise of the intracellular Ca^{2+} level and activation of the insulin secreting system.

Recent studies indicate that sulphonylureas bind to a specific receptor on the plasma membrane of the B cell and block ATP-dependent K^+ channels, and resultant depolarization initiates chain events leading to the secretion of insulin [4, 5].

It is of interest, therefore, to know the change of the K^+ flux, an initial phenomenon responding to secretagogues, which may be closely related to the stimulus recognition step in the pancreatic B cell. An in vitro study using isolated pancreatic islets of mice was performed to investigate the change of the K^+ efflux after the loading of glucose or glibenclamide, a potent hypoglycemic sulphonylurea.

Methods

Male NON mice and non-diabetic DDY mice weighing 20–25 g, maintained in a temperature-controlled room and fed ad libitum, were used. Two hours prior to the operation, the animals were injected intraperitoneally with 0.2 ml of 4% pilocarpine hydrochloride. The pancreatic islets of Langerhans were isolated by a collagenase-digestion method [6]. Whole procedures were performed at ice-cold temperature, except for a period of collagenase digestion and incubation. About 3 800 units of collagenase (Type IV Worthington Biochemical Co.) dissolved in 4 ml of Krebs-Ringer-Bibarconate Buffer (KRBB), pH 7.4, was injected via common bile duct. After ligation of common bile duct, pancreas was kept at 37°C for about 20 min. The collagenase digestion mixture was washed by cold KRBB and layered on the top of Ficoll-Hypacque solution. After the centrifugation at 2000 g for 20 min., islets were harvested and washed by KRBB. For insulin secretion study, a batch incubation was done. Five islets for each batch were placed in siliconized glass vials and preincubated for 30 min. with 0.5 ml of KRBB containing 3.3 mM glucose and 5 mg/ml bovine serum albumin (Fraction V Armour Pharmaceutical Co.), under the atmosphere of 95% O_2 and 5% CO_2 at 37°C. The islets were then transferred to fresh KRBB containing various substances and incubated for another 30 min. An aliquot of the medium was removed and immunoreactive insulin were assayed by radioimmunoassay, using rat insulin as the standard [6]. The ionic constituent (mM) of the standard KRBB was Na^+ 144.0, K^+ 5.9, Cl^- 128.0, Ca^{2+} 2.5, Mg^{2+} 1.2, HCO_3 25.0, SO_4^{2-} 1.2, H_2PO_4 1.2. As secretagogues, 20 mM L-arginine, 10 μM forskolin, an activator of adenylate cyclase, or 10 μM BAY K-8644, an activator of voltage-dependent calcium channel were used.

Perfusion system to monitor the efflux of $^{86}Rb^+$ used as tracer of K^+ was performed by the method of Henquin [3]. One hundred islets were packed into a plastic chamber with 200 ml-volume and KRBB (5 mg/ml bovine serum albumin, 3.3 mM glucose) was perfused by using a peristaltic pump at a flow rate of 0.5 ml/min. Perfusate was collected every 2 min. and radioactivity was counted. To compare NON and DDY groups, data were expressed as 'relative' $^{86}Rb^+$ efflux to the efflux rate at 30th min., just before the induction of high glucose or glibenclamide.

Results

Insulin Release: In non-diabetic DDY mice, insulin release from isolated pancreatic islets responded to varying glucose concentration, 3.3, 8.3 and 16.7 mM (8.9 ± 1.2, 16.8 ± 2.1, 22.7 ± 3.3 mU/islet/30 min., respectively). In NON mice, insulin release from the islet did not respond to high glucose (8.3 and 16.7 mM) (Fig. 1). However, insulin release from NON mice islets potentiated by 20 mM arginine, 10 μM fors-

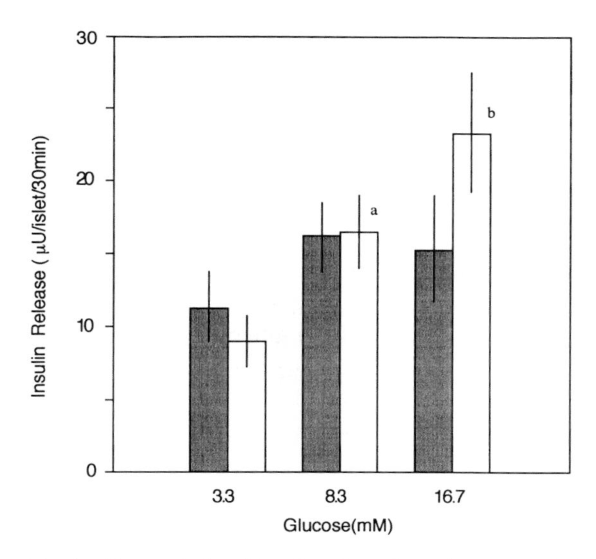

Fig. 1. Glucose-induced insulin release from the pancreatic islet of NON mice (shaded column) and control DDY mice (open column). Means ± SEM are shown. Letters on the top of columns indicate significant differences as compared with the value at 3.3 mM glucose in each animal group. a:$p < 0.05$, b: $p < 0.02$.

kolin, or 10 μM BAY K-8644 (33.7 ± 2.5, 27.6 ± 5.7, 21.3 ± 3.6 mU/islet/30 min., respectively) were higher than insulin release stimulated by 8.3 mM glucose alone (16.7 ± 2.1 mU/islet/30 min) (Fig. 2). The islet insulin content of NON mice was not

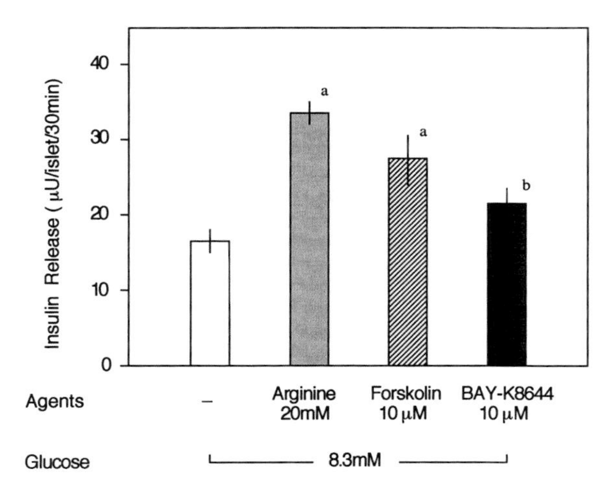

Fig. 2. Insulin release from the islet of NON mice stimulated by L-arginine, forskolin and BAY K-8644. Letters on the top of columns indicate significant differences from the value at 8.3 mM glucose alone. a: $p < 0.02$, b: $p < 0.05$.

decreased as compared with control mice (5.11±0.21 versus 4.57±0.29 mU/islet, respectively).

$^{86}Rb^{+}$ efflux: A progressive decrease in $^{86}Rb^{+}$ efflux recorded in the presence of 3.3 mM glucose (Fig. 3). The rate of $^{86}Rb^{+}$ efflux from perfused islets decline slowly and regularly with time after the addition of 250 nM glibenclamide. The efflux rate of both groups, DDY and NON mice, were quite comparable throughout the perfusion period (Fig. 3). Change of glucose concentration from 3.3 to 16.7 mM resulted in an augmented reduction of the efflux rate in normal islets (Fig. 4). In the islet of NON mice, the reduction of $^{86}Rb^{+}$ efflux rate in response to high glucose was blunted as compared to DDY mice.

Discussion

In the present study, the islet of NON mice, a mice model of NIDDM, exhibited a selective defect to secrete insulin responding to glucose. Because arginine, forskolin, or BAY K-8644 were able to potentiate insulin release in NON mice, signal transduction system such as adenylate cyclase, and voltage-dependent calcium channel except glucose recognition system might be intact. There is good agreement on that NIDDM is associated with a dysfunction of the B cell, which is an insensitivity of B cells to glucose. There are unsettled questions whether the B cell dysfunction is intrinsic to the B cell or secondary to the metabolic changes due to chronic hypergly-

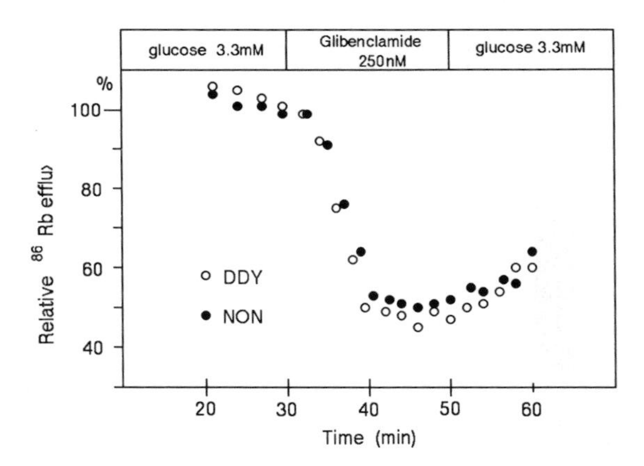

Fig. 3. $^{86}Rb^{+}$ efflux from the perfused islet of NON mice (closed circle) and DDY mice (open circle). Data are expressed as 'relative' $^{86}Rb^{+}$ efflux as compared with the efflux rate at 30th min., just before the induction of glibenclamide. Efflux rates of both group are not significantly different each other in whole perfusion period.

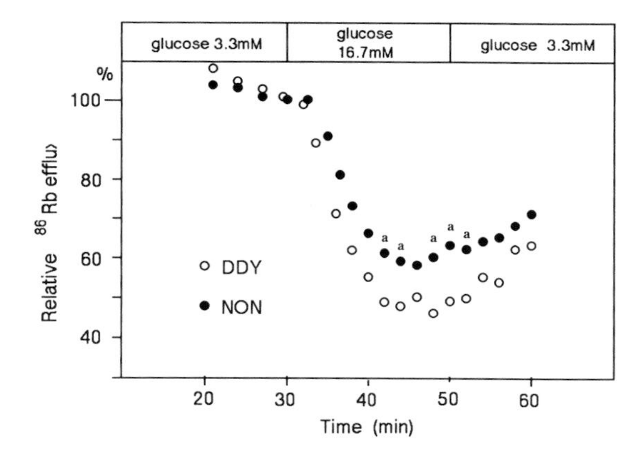

Fig. 4. $^{86}Rb^+$ efflux from the perfused islet of NON mice (closed circle) and DDY mice (open circle). Data are expressed as 'relative' $^{86}Rb^+$ efflux as compared with the efflux rate at 30th min., just before the induction of 16.7 mM glucose. Letters on the top of points indicate significant differences as compared with the value at 3.3 mM glucose in each animal group. a:$p<0.05$.

cemia. Intrinsic defects in the B cell include an inability to recognize glucose (glucoreceptor defect or impairment of glucose metabolism in B cells) [7], coupling defects between glucose stimulus and insulin secretion [2] and structural changes in the insulin secreting system [2], all of which result in abnormal insulin secretion.

According to the 'glucose metabolism hypothesis', insulin secretion is postulated to proceed as follows [8]. When ambient glucose levels rise, the B cell responds to the increase by metabolizing glucose and increasing intracellular levels of ATP. High intracellular ATP inhibits the ATP-sensitive K-channels and this depolarizes the B cell, causing the activation of voltage-sensitive Ca^{2+} channels, the influx of Ca^{2+} and the triggering of insulin release. The activity of the ATP-sensitive K channel is thought to play a central role in the physiological regulation of insulin secretion from the pancreatic B cell [8, 9]. Studies of rodent B cells and B cell lines have led to the proposal that an increase in the cytosolic ATP/ADP ratio, resulting from glucose metabolism, inhibits the channel activity. This produces membrane depolarization and precipitates a chain of events that culminates in insulin secretion. The channel is also the target for the hypoglycemic sulphonylureas used clinically in the treatment of NIDDM. These drugs inhibit channel activity and thereby stimulate electrical activity and insulin release [4, 5, 8, 9]. It has therefore been suggested that a defect in the regulation of the ATP-sensitive K channel may be involved in the initial deterioration of NIDDM [10].

The present study demonstrated that $^{86}Rb^+$ efflux evoked by glibenclamide was not disturbed in NON mice, although glucose-induced reduction of $^{86}Rb^+$ efflux was blunted. This clearly indicates that in NON mice, the B cell K channel responsible for glibenclamide remained intact, while the channel function in charge of glucose-induced membrane depolarization was altered. This raises the possibility that the ability of sulphonylureas to close the ATP-sensitive K-channels and subsequent stimulation of insulin release in sulphonylurea-sensitive NIDDM patients remain intact [10]. It can be postulated that the defect of NIDDM in the secretory function of the B cell, which is a glucose-insensitivity in insulin release, may be localized in either glucose recognition process including glucose metabolism or in the coupling factor between glucose recognition and the K channel. Involvement of other non-ATP-sensitive type of K-channels should not be excluded at present.

Acknowledgements

We are extremely grateful to Dr. S. Makino, Aburahi Laboratories, Shionogi Pharmaceutical Co., for supplying NON mice and useful suggestions. We are grateful to Ms F Inagaki for her technical assistance.

References

1 Makino S, Kunimoto K, Muraoka Y, Mizushima Y, Katagiri K, Tochino Y. Breeding of a nonobese, diabetic strain of mice. Exp Anim 1980;29:1–13.
2 Lacy PE. The physiology of insulin release. In: Volk BW, Wellman KF, eds. The diabetic pancreas. New York: Plenum Press, 1977;211–230.
3 Henquin JC. D-Glucose inhibits potassium efflux from pancreatic islet cells. Nature (London) 1978;271:271–273.
4 Gylfe E, Hellman B, Sehlin J, Taljedal I-B. Interaction of sulphonylurea with the pancreatic B-cell. Experientia 1984;40:1126–1134.
5 Sturgess NC, Ashford MLJ, Cook DL, Hales CN. The sulphonylurea receptor may be an ATP-sensitive potassium channel. Lancet 1985;1:474–475.
6 Taminato T, Seino Y, Goto Y, Imura H. Interaction of somatostati and calcium in regulating insulin release from isolated pancreatic islets of rats. Biochem Biophys Res Commun 1975;66(3);928–934.
7 Zawalich WS. Intermediary metabolism and insulin secretion from isolated ratislets of Langerhans. Diabetes 1979;28:252–262.
8 Ashcroft FM. Adenosine-5′-triphosphate-sensitive K^+ channels. Ann Rev Neurosci 1988;11:97–118.
9 Petersen OH, Findlay I. Electrophysilogy of the pancreas. Physiol Rev 1987;67:1054–1116.
10 Ashcroft FM, Kakei M, Kelly RP, Sutton R. ATP-sensitive K-channels in isolated human pancreatic β-cells. FEBS Lett 1987;215:9–12.

CHAPTER 8

Lipoproteins, cholesterol and bile acids in the NON mouse

KIYOHISA UCHIDA[a], HIROTSUNE IGIMI[a], HARUTO TAKASE[a], YASUHARU NOMURA[a], TOSHIYUKI CHIKAI[a], SUSUMU MAKINO[b], YOSHIYUKI HAYASHI[b], TAKAYUKI KAYAHARA[c], KAZUYA HIGASHINO[c] and NOZOMU TAKEUCHI[d]

[a]*Shionogi Research Laboratories, Fukushima-ku, Osaka, Japan*
[b]*Shionogi Aburahi Laboratories, Koga, Shiga, Japan*
[c]*The third Department of Internal Medicine, Hyogo College of Medicine, Nishinomiya, Japan*
[d]*The Central Laboratories, Ehime University Hospital, Shigenobu-cho, Ehime, Japan*

Current concepts of a new animal model: The NON mouse
Edited by N. Sakamoto, N. Hotta and K. Uchida

Contents

Introduction

A non-obese diabetic mouse (NOD) was established as an animal model for insulin-dependent diabetes mellitus [1]. The NOD mouse also shows a marked change in bile acid metabolism, an increase in cholic acid synthesis with a concomitant decrease of β-muricholic acid and a subsequent change in cholesterol metabolism that coincides with the appearance of glucosuria [2].

Strains of mice related to the NOD mouse are maintained by the Makino group [3]. A strain called NON first appeared as a control line for the NOD mouse, but now is regarded as an animal model for insulin-independent diabetes mellitus with disorders in glucose tolerance [4].

In the present study, we examined the serum lipoproteins and the metabolism of cholesterol and bile acid of NON mice in comparison with ICR mice.

Materials and methods

Animals

Male and female NON and ICR mice were bred in our laboratory. Ten mice at the ages of 3, 6, 9 and 12 months were individually housed in metabolic cages and kept in an air-conditioned room (25 ± 1°C, 50–60% humidity) lit 12 h a day (8.00 a.m. to 8.00 p.m.) with free access to chow diet (Japan Clea, CA-1) and tap water. Feces were collected for two days before the killing.

Before autopsy, the animals were made to fast for 5 h and the gallbladder was removed under sodium pentobarbital anesthesia (50 mg/kg, i.p.). Next, blood was withdrawn from the abdominal aorta with a heparinized syringe and the liver was removed.

Serum and liver lipid determination

Blood was kept for at least 1 h at room temperature, and the serum was obtained by centrifugation at 3000 rpm for 15 min. About 1 g of the largest lobe of the liver (lobus sinistra externa) was excised and homogenized with 9 vol of ice-chilled water using an Ultra-Turrax TP 18–10 (IKE-WERK, Janke and Kunkel KG, FRG). A portion of the homogenate (usually 1 ml) was lyophilized. The serum and the liver preparations were extracted with 10 ml of ethanol by refluxing for 20 min at 90°C. After filtration, portions of the extracts were evaporated to dryness under a stream of nitrogen, and the residues were dissolved in isopropyl alcohol.

Serum total and free cholesterol [5], phospholipids [6] and triglycerides [7] were determined by enzymatic colorimetric tests, and HDL-, HDL_2 and HDL_3-cholesterol levels were determined by the method of Gidez et al. [8].

A portion (0.25 ml) of the individual serum was combined by group, and the serum lipoproteins were separated by ultracentrifugation as described by Havel et al. [9] to obtain chylomicron, VLDL ($d<1.006$), IDL ($1.006<d<1.020$), LDL ($1.020<d<1.064$) and HDL ($1.064<d<1.21$) fractions. The cholesterol, phospholipid and triglyceride levels were determined by enzymatic methods [5–7], and the protein level was determined by the method of Lowry et al. [10]. Liver lipid levels were determined as reported previously [11].

Biliary lipid determination

The gallbladder was crushed in 20 ml ethanol with a glass rod and biliary lipids were extracted by refluxing for 10 min. at 80–90°C. After filtration, a portion was evaporated to dryness under a stream of nitrogen and the residue was hydrolyzed using cholylglycine hydrolase (EC 3.5.1.24, Sigma) [12]. Cholesterol was extracted with diethyl ether and then the bile acids were extracted with diethyl ether after the mixture had been acidified with 2 N hydrochloric acid solution. Bile acids were methylated with freshly prepared diazomethane and then trifluoroacetylated with trifluoroacetic anhydride [13]. The bile acid derivatives were quantified by gas-liquid chromatography utilizing a Hewlett Packard gas-chromatograph Model HP5890A equipped with a hydrogen flame ionization detector. A 15 m × 0.25 mm I.D. capillary column coated with DB-17 (J&W Scientific) was used. The operation temperature was 200–280°C at a rate of 5°C/min. for the column, 280°C for the injection port and 300°C for the detector. When sterols were analyzed by GLC, a 25 m × 0.32 mm I.D. capillary column coated with ULTRA Performance 19091A-112 (Hewlett Packard) was used at 300°C for the column, injection port and detector. Biliary phospholipids were determined by the method of Gomori [14].

Fecal sterol and bile acid determinations

Fecal sterols and bile acids were determined as reported previously [13, 15], but the hydrolysis of bile acids was performed using cholylglycine hydrolase [12] and the quantification of sterols and bile acids using capillary columns for GLC in the present study. Briefly, dried and powdered feces were extracted with absolute ethanol and petroleum ether, and hydrolyzed by cholylglycine hydrolase. After extraction of sterols with diethyl ether, the hydrolysate was acidified and bile acids were extracted with diethyl ether. The sterols and bile acids were quantified by GLC with capillary columns.

Statistical analysis

The results are expressed as the mean ± SEM. Analysis of variance was followed by Tukey's multiple comparison test and also by the Student's t-test.

Results and discussion

Body weight, liver weight, diet intake, serum and liver lipid levels, serum lipoproteins, biliary lipid levels, and fecal sterol and bile acid levels of NON and ICR mice aged 3, 6, 9 and 12 months are given in Tables I to VII. Apparent age-related

TABLE I
Body weight, liver weight, diet intake and feces dry weight in NON and ICR mice

	Age (months)	Male		Female	
		NON	ICR	NON	ICR
Body wt (g)	3	34.8 ±0.4[a]	40.7 ±0.6	31.4 ±0.7	33.1 ±0.7
	6	35.6 ±0.7	42.3 ±1.0	34.2 ±1.3	36.4 ±0.8
	9	38.6 ±0.6	44.8 ±0.8	34.7 ±0.60	39.0 ±1.1
	12	38.8 ±0.5	43.2 ±0.6	34.4 ±0.8	38.1 ±1.0
Liver wt (g/10gBW)	3	0.38±0.01	0.46±0.02	0.38±0.01	0.44±0.01
	6	0.34±0.01	0.46±0.01	0.32±0.01	0.42±0.01
	9	0.35±0.01	0.47±0.01	0.31±0.01	0.42±0.01
	12	0.35±0.01	0.48±0.01	0.32±0.01	0.40±0.01
Diet intake (g/day)	3	5.4 ±0.3	5.3 ±0.6	4.0 ±0.3	5.1 ±0.3
	6	3.7 ±0.3	4.8 ±0.5	3.5 ±0.0	3.7 ±0.3
	9	5.1 ±0.3	5.2 ±0.4	2.8 ±0.3	3.5 ±0.2
	12	4.3 ±0.4	5.0 ±0.5	2.4 ±0.2	3.6 ±0.2
Feces dry wt (g/day)	3	1.27±0.09	1.32±0.14	0.86±0.04	1.19±0.03
	6	0.81±0.10	1.15±0.09	0.75±0.02	0.83±0.05
	9	1.16±0.07	1.21±0.08	0.66±0.06	0.82±0.07
	12	0.94±0.08	1.17±0.12	0.62±0.05	0.84±0.02

[a]Mean ± SE in 10 mice.

changes were found in the body weight and gallbladder bile weight, but most of the determinants did not seem to show age-related changes. Therefore, the individual data for different ages were classified by group without regard to age, and the values were mostly compared between the strains or sexes. The HDL_2-cholesterol/HDL_3-cholesterol ratio seemed to increase with age in NON mice but not in ICR mice.

As shown in Fig. 1, the body weight of NON mice was less than that of ICR mice for both males and females. Similar but more marked results were obtained for the liver weight.

The serum cholesterol levels were low in the ICR female mice but almost the same for the other three groups. The liver cholesterol levels were higher in female mice of both strains due to the higher esterified cholesterol levels. The serum phospholipid levels were higher in NON mice than in ICR mice, and in males than in females. The liver phospholipid levels were not consistent with the serum phospholipid levels. The NON male mice showed extremely high levels of liver phospholipids. The triglyceride levels were rather constant in both serum and liver, except that the NON male mice showed higher serum triglyceride levels and the ICR female mice higher liver triglyceride levels.

Lipoprotein cholesterol levels are shown in Fig. 2. HDL-cholesterol levels were higher in males than in females but no significant difference was found between the two strains of mice. In addition, the HDL_2-cholesterol levels as well as the ratios of HDL_2-cholesterol/HDL_3-cholesterol were higher in male mice. As for the

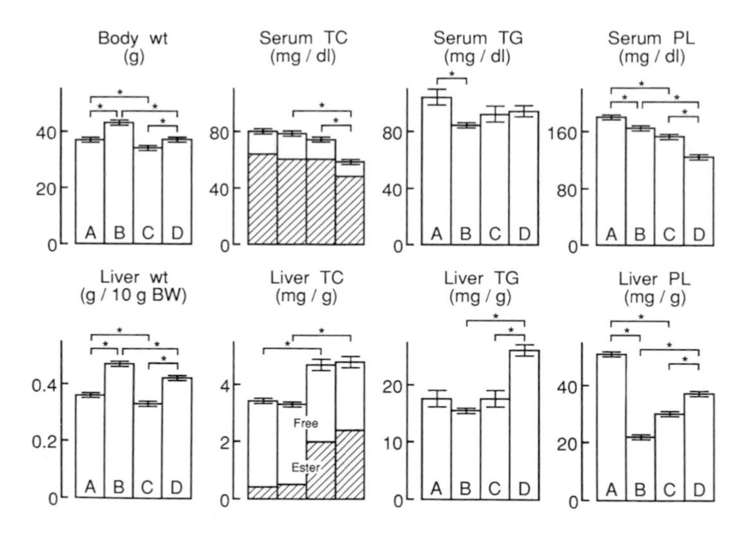

Fig. 1. Body weight, liver weight and serum and liver lipid levels in NON and ICR mice. The columns and bars show mean ± SE in 40 mice. A: NON male, B: ICR male, C: NON female, D: ICR female, and $*P < 0.05$.

TABLE II
Serum cholesterol, phospholipid and triglyceride levels in NON and ICR mice

	Age (months)	Male		Female	
		NON	ICR	NON	ICR
Total cholesterol (mg/dl)	3	82.3± 2.1[a]	85.1±5.0	78.6±2.2	71.0±4.9
	6	71.7± 2.4	74.9±2.9	64.3±1.8	58.2±2.4
	9	75.5± 3.4	77.7±3.9	70.9±2.8	52.9±4.8
	12	88.6± 5.0	72.5±3.7	78.1±2.5	52.0±3.9
Ester cholesterol (%)	3	78.9± 0.8	78.7±0.4	80.1±0.5	82.1±0.4
	6	81.8± 0.8	79.4±0.4	83.8±0.7	83.1±0.4
	9	79.6± 0.9	78.7±0.5	82.7±0.7	82.7±0.6
	12	78.7± 1.0	77.2±1.1	79.6±1.0	84.2±0.8
Phospholipids (mg/dl)	3	193.3± 5.2	171.5±7.8	160.3±3.7	143.5±7.3
	6	160.3± 4.5	157.6±3.9	133.1±4.7	119.7±3.0
	9	174.8± 5.0	171.3±6.4	151.9±4.7	115.5±6.8
	12	195.7± 7.9	158.5±6.2	155.8±4.0	116.0±6.0
Triglycerides (mg/dl)	3	143.5±10.5	86.6±4.9	138.1±8.1	108.4±9.3
	6	83.8± 5.4	93.0±5.3	78.5±5.0	86.4±4.3
	9	91.7± 7.6	86.4±4.2	74.4±6.5	88.5±6.4
	12	98.9± 6.1	73.5±3.6	73.6±6.4	85.0±4.0

[a]Mean ± SE in 10 mice.

(VLDL + LDL)-cholesterol levels, extremely high values were found in the NON female mice, while the ICR female mice showed low values. The ratios of (VLDL + LDL)-cholesterol/HDL-cholesterol were higher in female mice, especially in NON female mice, than in the males.

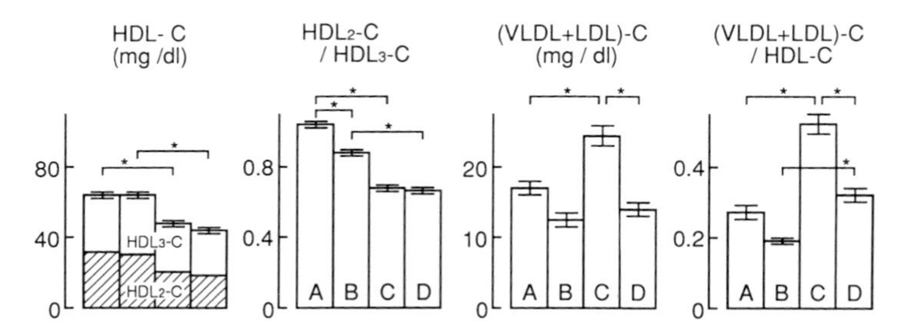

Fig. 2. Serum lipoprotein cholesterol levels in NON and ICR mice. The columns and bars show mean ± SE in 40 mice. A: NON male, B: ICR male, C: NON female, D: ICR female, and $*P < 0.05$.

TABLE III
HDL, VLDL and LDL cholesterol (C) levels in NON and ICR mice

	Age (months)	Male		Female	
		NON	ICR	NON	ICR
HDL-C (mg/dl)	3	65.0 ±2.4[a]	69.5 ±2.9	48.6 ±1.7	54.4 ±3.2
	6	57.3 ±2.7	62.5 ±1.9	45.3 ±3.0	41.4 ±1.9
	9	63.9 ±2.3	66.6 ±1.2	49.4 ±4.0	40.0 ±4.5
	12	66.9 ±2.4	61.1 ±1.7	49.8 ±1.7	42.5 ±3.9
HDL_2-C/HDL_3-C ratio	3	0.83±0.02	0.79±0.08	0.54±0.04	0.63±0.04
	6	1.00±0.14	0.80±0.06	0.64±0.07	0.62±0.03
	9	1.12±0.07	0.77±0.05	0.79±0.06	0.63±0.07
	12	1.18±0.07	1.11±0.11	0.74±0.11	0.72±0.04
(VLDL+LDL)-C (mg/dl)	3	17.3 ±1.9	15.7 ±2.3	30.0 ±2.3	16.7 ±1.5
	6	14.4 ±2.2	12.5 ±1.4	19.1 ±0.9	16.8 ±1.1
	9	13.9 ±1.9	11.1 ±1.8	21.5 ±2.1	12.9 ±2.1
	12	21.7 ±2.7	11.4 ±2.2	28.3 ±1.7	9.6 ±1.4
(VLDL+LDL)-C/HDL-C ratio	3	0.27±0.03	0.22±0.03	0.62±0.06	0.31±0.02
	6	0.26±0.05	0.20±0.02	0.43±0.05	0.41±0.02
	9	0.22±0.03	0.17±0.03	0.46±0.09	0.33±0.05
	12	0.33±0.04	0.19±0.04	0.57±0.04	0.23±0.03

[a]Mean ± SE in 10 mice.

As shown in Table IV, the serum chylomicron levels were higher and the VLDL levels were lower in NON mice than in ICR mice. The IDL, LDL and HDL levels fluctuated somewhat but no consistent difference was found between the strains or sexes. The total lipoprotein levels were higher in males than in females.

The VLDL levels of NON mice, both male and female, were high at the age of 3 months but decreased to less than half of the initial ones, after 6 months of age. Such an age-related change, though to a much lesser extent, was also found in ICR female mice, but not in ICR males. The composition ratios of protein, cholesterol, triglycerides and phospholipids, however, were almost the same for the young (3 months old) and aged mice (6–12 months old) as shown in Fig. 3.

The gallbladder bile weight showed an age-related change in all groups (Table VI), with the youngest animals showing the lowest values. No significant correlation was found between the gallbladder bile weight and liver weight (Fig. 4). The biliary cholesterol, phospholipid and bile acid levels are summarized in Fig. 5. NON male mice showed higher values for these lipid levels than ICR male mice, but such a significant strain-linked difference was not found in female mice. The biliary cholesterol levels

TABLE IV
Serum lipoprotein levels in NON and ICR mice

	Age (months)	Male		Female	
		NON	ICR	NON	ICR
Chylomicron (mg/dl)	3	39.3	10.4	30.7	15.6
	6	25.7	11.5	22.1	11.6
	9	22.5	10.5	25.4	14.5
	12	20.7	18.4	40.3	8.8
VLDL (mg/dl)	3	93.7	76.9	99.3	94.1
	6	46.4	80.5	41.9	70.4
	9	50.2	70.2	37.7	76.2
	12	61.4	58.6	33.9	66.3
IDL (mg/dl)	3	31.3	17.9	26.6	18.6
	6	20.6	15.5	16.6	19.6
	9	20.7	15.8	17.7	18.4
	12	26.9	16.4	18.2	18.7
LDL (mg/dl)	3	27.1	29.0	35.6	33.0
	6	29.0	21.6	33.7	26.0
	9	29.6	26.7	37.9	26.6
	12	44.7	32.5	40.7	23.9
HDL (mg/dl)	3	269.3	308.2	150.8	156.4
	6	226.4	231.7	132.8	128.0
	9	251.2	256.2	153.5	124.7
	12	265.7	219.9	152.3	135.1
Total (mg/dl)	3	460.7	442.4	343.0	317.7
	6	348.1	360.8	247.0	255.6
	9	374.2	379.4	272.2	260.4
	12	419.4	345.8	285.4	252.8

were lower in female mice than in males but the opposite was found for the phospholipid and bile acid levels.

The feces dry weight fluctuated somewhat with age and group, but was positively correlated to the diet intake (Fig. 6). The feces dry weight and fecal sterol and bile acid levels were summarized in Fig. 7. Generally, the feces dry weight was lower in NON mice than in ICR mice, and also lower in female mice than in males.

The fecal excretion of sterols was higher and that of bile acids was lower in NON male mice than in NON female mice, while such a sex-linked difference was not

TABLE V
Liver cholesterol, phospholipid and triglyceride levels in NON and ICR mice

	Age (months)	Male		Female	
		NON	ICR	NON	ICR
Total cholesterol (mg/g)	3	3.28±0.14[a]	3.39±0.22	4.20±0.23	4.74±0.34
	6	3.55±0.11	3.04±0.21	5.22±0.36	5.07±0.42
	9	3.16±0.13	3.13±0.13	4.51±0.50	4.47±0.17
	12	3.73±0.24	3.76±0.23	5.00±0.37	5.10±0.56
Ester cholesterol (mg/g)	3	0.36±0.05	0.55±0.13	1.58±0.17	2.25±0.29
	6	0.34±0.05	0.41±0.13	2.43±0.33	2.70±0.39
	9	0.24±0.05	0.42±0.11	1.88±0.42	1.99±0.15
	12	0.56±0.16	0.74±0.18	1.95±0.32	2.69±0.45
Phospholipids (mg/g)	3	51.7 ±1.9	24.6 ±0.9	32.2 ±0.7	39.3 ±0.3
	6	52.2 ±2.7	19.5 ±1.0	30.8 ±0.9	35.8 ±0.7
	9	49.6 ±1.7	21.3 ±0.9	29.2 ±1.1	36.8 ±0.5
	12	49.3 ±1.7	21.2 ±0.8	28.5 ±0.4	35.9 ±0.3
Triglycerides (mg/g)	3	15.7 ±0.8	17.9 ±1.6	12.8 ±1.3	28.5 ±2.6
	6	16.7 ±1.0	16.1 ±1.7	19.2 ±2.9	23.0 ±2.1
	9	19.6 ±1.4	14.1 ±1.3	18.2 ±3.8	25.4 ±2.5
	12	12.0 ±1.1	13.6 ±0.8	19.1 ±1.7	26.3 ±2.4

[a]Mean ± SE in 10 mice.

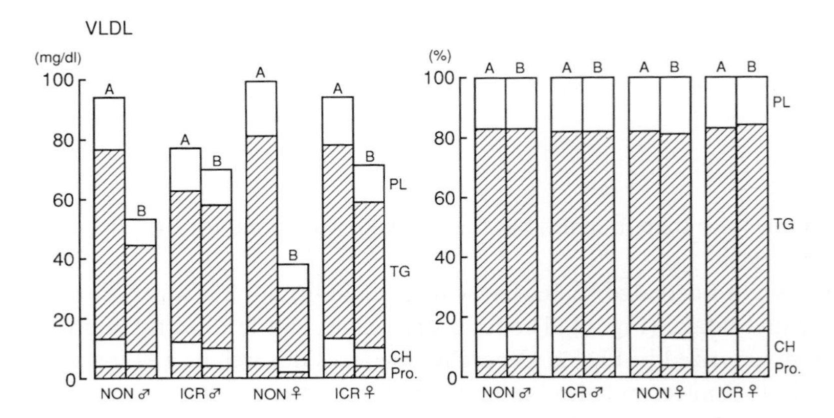

Fig. 3. Serum VLDL levels and lipid composition in VLDL fractions in NON and ICR mice. A: Values for mice aged 3 months. B: Values for mice aged 6–12 months. Pro: proteins, CH: cholesterol, TG: triglycerides, and PL: phospholipids.

TABLE VI
Gallbladder bile weight, biliary cholesterol, phospholipid and bile acid levels in NON and ICR mice

	Age (months)	Male		Female	
		NON	ICR	NON	ICR
Gallbladder bile (mg)	3	36.0 ±4.6[a]	31.2 ±3.9	25.0 ±3.3	30.6 ±3.6
	6	31.8 ±6.3	41.8 ±6.5	34.8 ±2.4	36.6 ±5.8
	9	42.0 ±3.6	45.2 ±5.0	51.8 ±2.1	59.0 ±6.1
	12	53.2 ±3.3	60.8 ±6.4	64.4 ±6.1	51.3 ±5.2
Cholesterol (mg/ml)	3	1.84±0.10	1.37±0.16	0.50±0.25	0.34±0.06
	6	2.20±0.50	1.19±0.11	0.22±0.08	0.29±0.03
	9	1.54±0.25	1.01±0.20	0.20±0.04	0.34±0.02
	12	1.41±0.25	1.03±0.15	0.24±0.05	0.40±0.12
Phospholipids (mg/ml)	3	13.4 ±0.7	10.8 ±1.2	21.2 ±2.1	20.0 ±1.8
	6	12.9 ±2.5	10.1 ±1.0	20.5 ±1.8	17.1 ±1.6
	9	13.0 ±1.7	8.4 ±0.3	20.5 ±2.3	13.0 ±2.8
	12	11.9 ±1.7	7.4 ±0.8	18.8 ±1.4	18.3 ±1.5
Bile acids (mg/ml)	3	37.8 ±5.3	38.2 ±3.3	61.2 ±4.6	51.8 ±6.2
	6	32.1 ±2.6	14.1 ±1.6	46.4 ±6.4	54.9 ±5.4
	9	27.9 ±5.6	23.1 ±6.8	44.3 ±2.2	41.9 ±3.9
	12	27.0 ±5.6	23.9 ±2.7	44.0 ±4.6	47.3 ±2.8
CA/CDCA ratio	3	1.1 ±0.2	0.9 ±0.1	1.6 ±0.1	1.1 ±0.1
	6	1.3 ±0.3	1.4 ±0.1	2.1 ±0.2	1.2 ±0.1
	9	1.8 ±0.3	1.3 ±0.2	2.4 ±0.2	1.4 ±0.1
	12	1.8 ±0.3	1.2 ±0.2	2.3 ±0.1	1.7 ±0.3

[a]Mean ± SE in 5 mice.

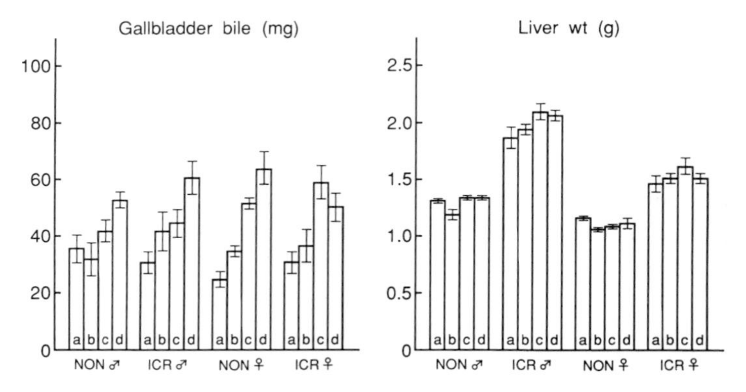

Fig. 4. Gallbladder bile weight and liver weight in NON and ICR mice. The columns and bars show mean ± SE in 10 mice., a: 3 months, b: 6 months, c: 9 months, and d: 12 months

TABLE VII
Fecal sterol and bile acid levels in NON and ICR mice

	Age (months)	Male		Female	
		NON	ICR	NON	ICR
Total sterols (mg/day)	3	4.97±0.32[a]	2.57±0.48	2.41±0.13	3.07±0.18
	6	2.69±0.49	2.11±0.17	1.87±0.12	2.13±0.17
	9	3.47±0.38	2.89±0.24	1.66±0.26	2.22±0.13
	12	2.40±0.32	2.87±0.43	1.26±0.09	1.76±0.14
Cholesterol (mg/day)	3	3.97±0.38	2.45±0.47	1.89±0.13	2.83±0.17
	6	2.12±0.41	1.82±0.14	1.52±0.14	1.97±0.17
	9	3.03±0.27	2.66±0.23	1.24±0.23	2.01±0.12
	12	2.07±0.32	2.66±0.41	0.94±0.06	1.63±0.13
Coprostanol (mg/day)	3	1.00±0.20	0.12±0.02	0.53±0.02	0.24±0.10
	6	0.57±0.08	0.29±0.08	0.35±0.04	0.15±0.03
	9	0.44±0.12	0.25±0.04	0.42±0.03	0.21±0.06
	12	0.35±0.07	0.21±0.03	0.32±0.05	0.14±0.01
Bile acids (mg/day)	3	1.84±0.20	2.45±0.22	2.53±0.14	3.57±0.34
	6	1.13±0.12	2.39±0.19	2.61±0.18	2.39±0.23
	9	1.68±0.08	2.97±0.14	2.54±0.25	2.56±0.30
	12	1.63±0.14	3.31±0.37	2.03±0.14	3.01±0.29
CA/CDCA ratio	3	1.3 ±0.1	0.09±0.1	0.9 ±0.0	1.0 ±0.1
	6	1.1 ±0.1	0.8 ±0.0	0.9 ±0.1	1.0 ±0.1
	9	1.0 ±0.1	1.1 ±0.1	0.8 ±0.1	1.1 ±0.1
	12	0.9 ±0.0	1.0 ±0.0	0.8 ±0.1	1.5 ±0.3

[a]Mean ± SE in 5 mice.

found in ICR mice. In addition, the coprostanol levels were higher in NON mice than in ICR mice of both sexes.

The biliary and fecal bile acid compositions are given in Tables VIII and IX. Cholic acid and β-muricholic acid are two major bile acids in mouse bile [16], and deoxycholic acid and ω-muricholic acid, found in the feces in larger amounts, are secondary bile acids formed by the action of intestinal bacteria from cholic acid and β-muricholic acid, respectively.

The composition ratios of cholic acid in the bile were about 10% higher in NON mice than in ICR mice in both sexes. The NON female mice showed low values in β-muricholic acid and high values in chenodeoxycholic acid, but such a trend was not found in ICR female mice.

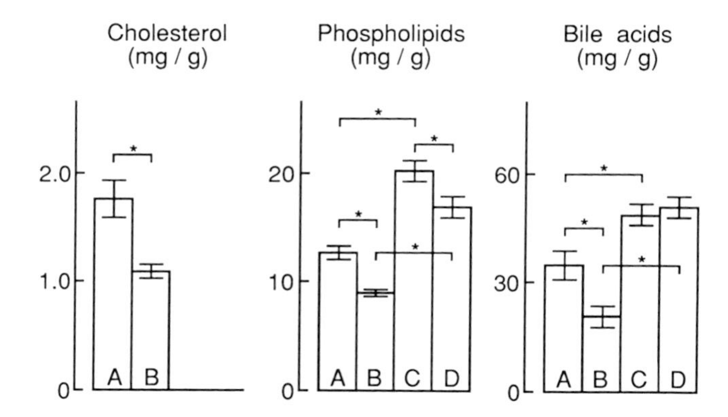

Fig. 5. Biliary lipid levels in NON and ICR mice. The columns and bars show mean ± SE in 20 mice. A: NON male, B: ICR male, C: NON female, D: ICR female, and $*P < 0.05$.

TABLE VIII
Biliary bile acid composition in NON and ICR mice

	Male		Female	
	NON	ICR	NON	ICR
Cholic acid group (%)				
Cholic acid	53.2±2.4[a]	43.9±1.6	57.2±1.1	48.8±1.5
Deoxycholic acid	2.2±0.1	3.2±0.3	4.0±0.1	2.5±0.1
3α, 12=0	1.8±0.2	9.8±1.8	8.9±0.4	6.2±0.2
Chenodeoxycholic acid group (%)				
Chenodeoxycholic acid	1.4±0.1	0.7±0.2	4.1±0.3	2.1±0.1
Ursodeoxycholic acid	1.1±0.1	1.2±0.1	0.9±0.0	1.0±0.0
Hyodeoxycholic acid	1.5±0.2	1.9±0.2	1.9±0.3	1.9±0.2
α-Muricholic acid	4.1±0.3	4.9±0.7	3.9±0.2	3.6±0.2
β-Muricholic acid	24.8±2.1	27.7±1.9	15.3±1.0	28.7±1.5
ω-Muricholic acid	8.2±0.8	4.9±0.3	2.6±0.1	2.7±0.1
Lithocholic acid	0.4±0.0	0.3±0.0	0.4±0.0	0.5±0.0
3α, 7=0	0.4±0.1	0.2±0.0	0.8±0.2	0.3±0.0
3α, 6=0	nd	nd	0.4±0.1	nd
Others (%)	1.8±0.2	9.8±1.8	8.9±0.4	6.2±0.2

[a]Mean ± SE in 20 mice aged 3–12 months.

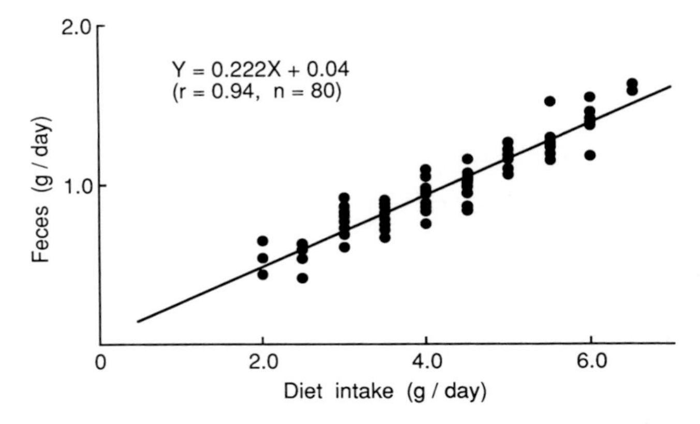

Fig. 6. Relationship between diet intake and feces dry weight in NON and ICR mice.

In the feces, the composition ratios of cholic acid were almost the same for all the groups, but those of deoxycholic acid were higher in females and those of 3α, 12=0 (3α-hydroxy-12-oxo-5β-cholanoic acid) were higher in males. As for the relationship of these bile acids to cholic acid, a significant difference was found between sexes but not between strains. However, the composition ratios of β-muricholic acid were higher in ICR mice and those of ω-muricholic acid were higher in NON mice. In addition, α-muricholic acid was higher in female mice.

TABLE IX
Fecal bile acid composition in NON and ICR mice

	Male		Female	
	NON	ICR	NON	ICR
Cholic acid group (%)				
Cholic acid	9.4±0.6[a]	7.7±0.7	9.1±0.5	11.4±1.0
Deoxycholic acid	21.8±0.6	21.4±0.7	26.2±0.9	26.7±1.5
3α, 12α, 7=0	2.5±0.3	3.0±0.3	2.6±0.2	2.4±0.4
3α, 12=0	17.8±1.8	17.2±1.2	8.5±0.5	11.2±0.4
Chenodeoxycholic acid group (%)				
α-Muricholic acid	2.3±0.1	2.9±0.1	8.5±0.4	5.1±0.3
β-Muricholic acid	11.0±0.5	22.8±0.7	14.6±1.3	20.9±1.2
ω-Muricholic acid	35.0±1.4	23.0±0.7	30.7±1.2	22.6±1.1
Lithocholic acid	n.d.	2.0±0.4	1.1±0.1	n.d.

[a]Mean ± SE in 20 mice aged 3–12 months.

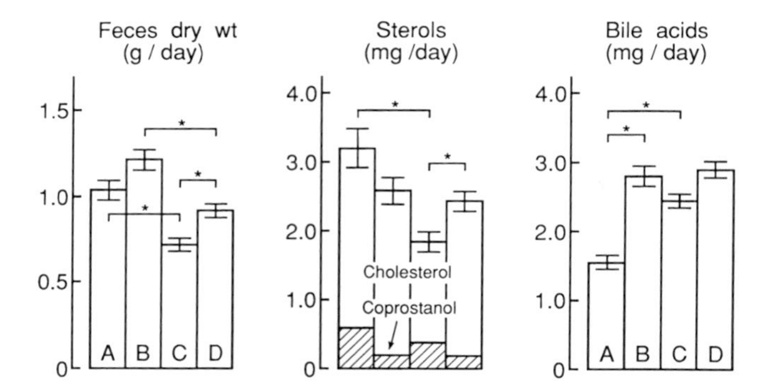

Fig. 7. Feces dry weight, and fecal sterols and bile acids in NON and ICR mice. The columns and bars show mean ± SE in 20 mice. A: NON male, B: ICR male, C: NON female, D: ICR female, and $*P < 0.05$.

In previous papers [15, 17], we reported that cholic acid increased with age while chenodeoxycholic acid (β-muricholic acid in rats and mice) decreased, resulting in an increase in the ratio of the cholic acid group to the chenodeoxycholic acid group (CA/CDCA). The composition ratio of cholic acid also increases in insulin-dependent diabetic animals such as NOD mice [2] or alloxan-treated mice [16] and rats [18]. In these animals, cholic acid becomes the sole major constituent accounting for more than 90% of the total bile acids. The fact that the composition ratios of cholic acid in the present NON mice were higher than those in ICR mice seemd to suggest that the bile acid metabolism of the NON mice was similar to that of NOD mice, but the increase of cholic acid was only found in the bile and not in the feces. In addition, the CA/CDCA ratio increased in the bile of NON male and female mice and of ICR female mice (Table VI), but not in the feces of all the groups (Table VII). Therefore, it would be premature to conclude that a change in bile acid metabolism, i.e., an increase of cholic acid synthesis, is found in insulin-independent diabetic animals (NON) as in insulin-dependent diabetes (NOD).

Summary

Serum and liver lipid levels, serum lipoprotein levels, and biliary and fecal sterol and bile acids levels were measured for male and female NON and ICR mice aged 3, 6, 9 and 12 months. Age-related changes were only found in the body weight and gallbladder bile weight, and not for the other determinants. Some strain-linked or sex-linked differences were observed. NON mice, both male and female, had lower body weight and liver weight than ICR mice. The liver cholesterol levels, especially esterified cholesterol levels, were higher in the females. The serum chylomicron levels were higher but the VLDL levels were lower in NON mice. The HDL levels were high in the NON male mice. However, the lipid compositions in each lipoprotein fraction were almost the same for NON and ICR mice or for male and female mice. The fecal excretion of sterols was higher and that of bile acids was lower in NON male mice than ICR males, but such a difference was not found in NON female mice. No sex-linked difference was found for the ICR mice. The composition ratio of cholic acid in the bile was higher for the NON mice than the ICR mice, and that of β-murichoic acid was low only for NON female mice. In the feces, the deoxycholic acid level was lower and the 3α-hydroxy-7-oxo-5β-cholanoic acid level was higher for male mice, and the β-muricholic acid level was lower and the ω-muricholic acid level was higher in NON mice. The NOD mouse displays disturbance in its cholesterol and bile acid metabolism, but our findings suggest that a similar disturbance does not occur in the NON mouse.

Acknowledgements

We sincerely thank Misses Keiko Suzumura and Enatsu Komori of Shionogi Research Laboratories for their help in preparing the manuscript.

References

1 Makino S, Tochino Y. Spontaneously diabetic non-obese mice. Exp Anim 1978;27:27–29 (in Japanese).
2 Uchida K, Makino S, Akiyoshi T. Altered bile acid metabolism in nonobese, spontaneously diabetic (NOD) mice. Diabetes. 1985;34:79–83.
3 Makino S. The NOD mouse and its related strains. Clin Immunol 1990;22:989–993 (in Japanese).
4 Ohgaku S, Morioka H, Sawa T, Yano S, Yamamoto H, Okamoto H, Tochino Y. Reduced expression and restriction fragment length polymorphism of insulin gene in NON mice, a new animal model for nonobese, noninsulin-dependent diabetes. In: Shafrir E, Renold AE, eds. Frontiers in Diabetes Research, Lessons from Animal Diabetes II. John Libbey, 1988;319–323.
5 Richmond W. Preparation and properties of a cholesterol oxidase from Nocardia sp. and its application to the enzymatic assay of total cholesterol in serum. Clin Chem 1973;19:1350–1356.
6 Takayama M, Ito S, Nagasaki T, Tanimizu I. A new enzymatic method for the determination of serum choline-containing phospholipids. Clin Chim Acta 1977;79:93–98.
7 Bucolo G, Yabut J, Chang TY. Mechanized enzymatic determination of triglycerides in serum. Clin Chem 1975;21:420–424.
8 Gidez LI, Miller GJ, Burstein M, Slagle S, Eder HA. Separation and quantitation of subclasses of human plasma high density lipoproteins by a simple precipitation procedure. J Lipid Res 1982;23:1206–1223.
9 Havel RJ, Eder HA, Bragdon JH. The distribution and chemical composition of ultracentrifugally separated lipoproteins in human serum. J Clin Invest 1955;34:1345–1353.

10 Lowry OH, Rosebrough NJ, Farr AL, Randall RJ. Protein measurement with Folin phenol reagent. J Biol Chem 1951;193:265–276.
11 Takeuchi N, Itoh M, Yamamura Y. Esterification of cholesterol and hydrolysis of cholesteryl ester in alcohol induced fatty liver of rats. Lipids 1974;9:353–357.
12 Padmanabhan PN. A modified gas-liquid chromatographic procedure for the rapid determination of bile acids in biological fluids. Anal Biochem 1969;29:164–171.
13 Uchida K, Nomura Y, Kadowaki M, Takeuchi N, Yamamura Y. Effect of dietary cholesterol on cholesterol and bile acid metabolism in rats. Jpn J Pharmacol 1977;27:193–204.
14 Gomori G. A modification of the colorimetric phosphorus determination for use with the photoelectric colorimeter, J Lab Clin Med 1942;27:955–960.
15 Uchida K, Nomura Y, Kadowaki M, Takase H, Takano K, Takeuchi N. Age-related changes in cholesterol and bile acid metabolism in rats. J Lipid Res 1978;19:544–552.
16 Akiyoshi T, Uchida K, Takase H, Nomura Y, Takeuchi N. Cholesterol gallstones in alloxan-diabetic mice. J Lipid Res 1986;27:915–924.
17 Uchida K, Igimi H, Takase H, Nomura Y, Nakao H, Chikai T, Ichihashi T, Kayahara T, Takeuchi N. Bile acids and aging in relation to hypercholesterolemia in rats, In: Kitani K, ed. Liver and Ageing-1990. Amsterdam: Elsevier, 1991:153–164.
18 Uchida K, Takase H, Kadowaki M, Nomura Y, Matsubara T, Takeuchi N. Altered bile acid metabolism in alloxan diabetic rats. Jpn J Pharmacol 1979;29:553–562.

CHAPTER 9

Intranuclear filaments in the pancreatic B cells and abnormal glucose tolerance in the NON mouse

K. YAMAMOTO, J. MIYAGAWA, T. HANAFUSA and N. KONO

The Second Department of Internal Medicine, Osaka University Medical School, Osaka, Japan

Current concepts of a new animal model: The NON mouse
Edited by N. Sakamoto, N. Hotta and K. Uchida

Contents

Introduction

The non-obese non-diabetic (NON) mouse which derived from the Jcl-ICR mouse is a sister strain of the non-obese diabetic (NOD) mouse. The NON mouse has been reported to show glucose intolerance and hypoinsulinemia without overt diabetes [1–3]. Thus, the NON mouse is proposed to be an animal model of human non-obese, Type 2 (non-insulin-dependent) diabetes mellitus (NIDDM). So far, no pathologic findings in the islet of the NON mouse have been reported by light microscopy. Described here are the electron microscopic observations of islets of the NON mouse with special reference to the presence of insulitis and retrovirus-like particles and other abnormalities in B cells.

Materials and methods

Light microscopy

A total of 15 NON mice were perfused with Bouin's solution. The pancreas was then excised and immersed in the same solution for several hours. The pancreatic tissues were dehydrated and embedded in paraffin. Sections were cut on a microtome and stained with hematoxylin-eosin and observed by a light microscope.

Electron microscopy

A total of 28 NON mice were perfused with glutaraldehyde (3.0%). The islets were then excised from the pancreas and immersed in the same fixative for two hours. The animals included 4 male and 4 female 8-week-old, 4 male 17-week-old, 4 male and 4 female 26-week-old, and 4 male and 4 female 40-week-old mice. The islets were post-fixed with OsO_4 (1.0%) for one hour and block-stained with aqueous uranyl acetate solution for 45 min. The blocks were dehydrated in graded concentrations of ethanol series and embedded in epoxy resin. Ultrathin section cut on an ultramicrotome were doubly stained with aqueous uranyl acetate solution and Reynolds' lead citrate. The sections were then observed in a transmission electron microscope.

Results

Light microscopy

Islet cells showed no apparent morphologic abnormalities by light microscopic studies of pancreas sections stained with hematoxylin-eosin. There was no evidence of fibrosis, hyalinization, or abnormal pigmentation, which are morphologic characteristics in islets of animal models of Type 2 diabetes. Observation of semithin sections from the blocks fixed for electron microscopy and stained with toluidine blue revealed a cluster of lymphocytes in contact with islets in 2 male mice (26 and 40 weeks of age). However, in contrast to the NOD mouse [4], apparently lymphocytic infiltration into the islets ('insulitis') was not recognized.

Electron microscopy

Intranuclear filaments in pancraetic B cells were observed by electron microscopic studies (Fig. 1). Bundles consisting of many fine intranuclear filaments were seen only in B cells, and not in A or D cells. Other intracellular organelles such as Golgi com-

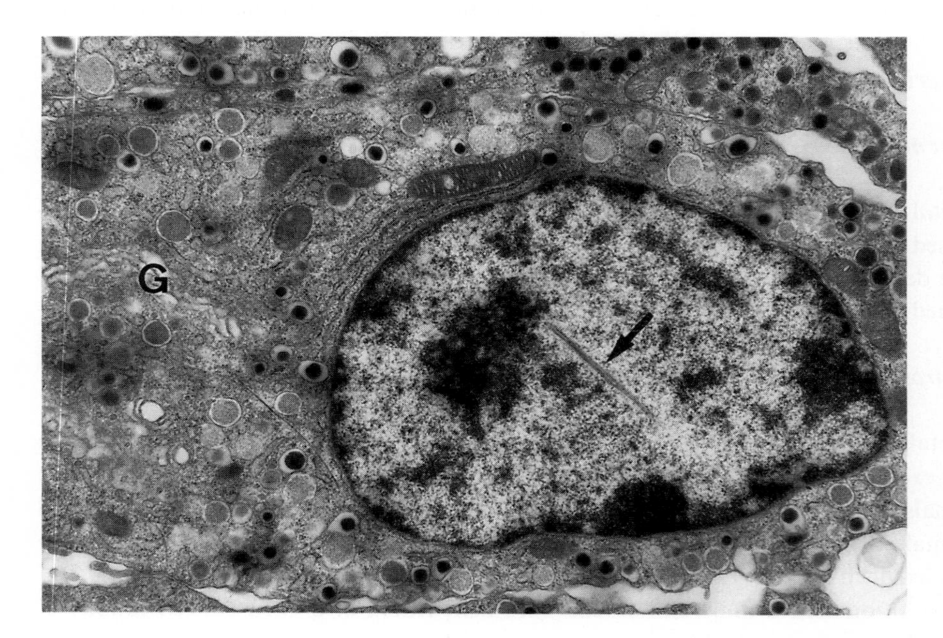

Fig. 1. Pancreatic B cell of 26-week-old female NON mouse. A bundle of intranuclear filaments in B cells (arrow) is observed. Other intracellular organelles appear to be normal. Retrovirus-like particles are not recognized. G: Golgi complex (× 9500).

plex, rough endoplasmic reticulum, and mitochondria appeared to be normal. There were no degenerative changes in B cells with intranuclear filaments, although the number of secretory granules seemed to be decreased in some B cells. Retrovirus-like particles which were observed in pancreatic B cells of NOD mice [5] were not recognized in B cells of any NON mice examined.

The bundles of intranuclear filaments were 300 to 2300 nm long and about 150 mm wide, while each filament was about 8 nm in diameter. The bundles tapered at the end, and there were no apparent connections between the end of the bundles and nucleoli or nuclear membranes. From a cross-sectional picture of the intranuclear bundle, it appeared to consist of about 100 filaments (Fig. 2).

Incidence of intranuclear filaments

The incidence of intranuclear filaments increased with age, but did not differ significantly between male and female mice.

Intraperitoneal glucose tolerance test (IPGTT)

IPGTT (1.0 mg/g × body weight) of NON mice revealed a significantly higher plasma glucose level in 15-week old male NON mice than age-matched male ICR mice at 60 min. after the glucose load, whereas no other significant difference was recog-

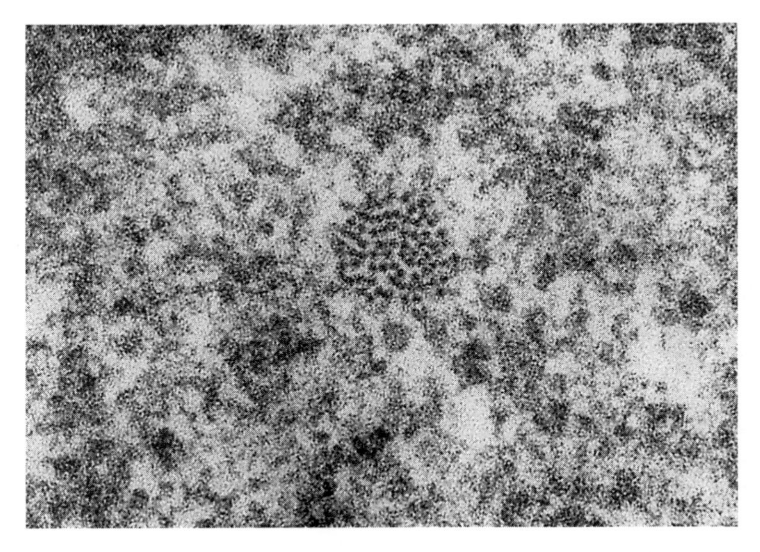

Fig. 2. Cross-sectional picture of the bundle. From this picture, the bundle consists of about 100 fine intranuclear filaments (× 130000).

nized between the glucose levels of NON mice and ICR mice irrespective of sex. Thus, male NON mice at 15 weeks of age showed abnormal glucose tolerance compared to age-matched ICR mice, but male mice of other age and female mice showed normal glucose tolerance.

Discussion

Morphologic appearance of pancreatic islet cells of NON mice were different from those of NOD mice, though both NON and NOD mice are animal models of diabetes and derived from ICR mice. Because there was no insulitis in islets of NON mice, glucose intolerance of these animals is supposed not to be related to the immune mechanism. On the other hand, bundles of intranuclear filaments were observed in pancreatic B cells of NON mice, and further investigation would clarify the relation of these filaments to glucose intolerance.

Intranuclear filaments have been reported to appear in various organs such as neurons of animals including humans, glial cells of hagfish and rabbit, and so on. These filaments are considered to be related to cellular activity, cell division or renewal, aging, and viral infection.

Intranuclear filaments in pancreatic insulin producing cells have been reported only in human insulinoma cells and in B cells of obese-hyperglycemic mice, both of which are known to synthesize and secrete insulin actively [6, 7]. Thus, the demonstration of the filaments in NON mouse B cells suggest that these cells are hyperactive due to some load on the cells.

References

1 Kano Y. Insulin secretion in NOD (non-obese diabetic) and NON (non-obese non-diabetic) mouse. J Kyoto Pref Univ Med 1988;97:295–308.

2 Funakawa S, Kanaya T, Ito H, Tochino Y. Glomerular changes in NON mice with long-term mild diabetes mellitus. Biomed Res 1988;9:409–412.

3 Sawa T, Ohgaku S, Morioka H, Yano S. Molecular cloning and DNA sequence analysis of preproinsulin genes in the NON mouse, an animal model of human non-obese, non-insulin-dependent diabetes mellitus. J Mol Endocrinol 1990;5:61–67.

4 Fujino-Kurihara H, Fujita H, Hakura A, Nonaka K, Tarui S. Morphological aspects on pancreatic islets of non-obese diabetic (NOD) mice. Virchows Arch B 1985;49:107–120.

5 Fujita H, Fujino H, Nonaka K, Tarui S, Tochino Y. Retrovirus-like particles in pancreatic B-cells of NOD (non-obese diabetic) mice. Biomed Res 1984;5:67–70.

6 Bencosme SA, Allen RA, Latta H. Functioning pancreatic islet cell tumors studied electron microscopically. Am J Pathol 1963;42:1–21.

7 Boquist L. Intranuclear rods in pancreatic islet β-cells. J Cell Biol 1969;43:377–381.

SECTION III

Characteristics of the nephropathy in the NON mouse

CHAPTER 10

Histopathological observation of the development of glomerular intracapillary deposits in the NON mouse

YOSHIHIRO MURAOKA[a], SHINJI MATSUI[a], HIROSHI WATANABE[a] and SUSUMU MAKINO[b]

[a]*Kanzakigawa Laboratory, Shionogi Research Laboratories, Shionogi & Co., Ltd., Toyonaka, Osaka, Japan*
[b]*Aburahi Laboratories, Shionogi & Co., Ltd., Koka, Shiga, Japan*

Current concepts of a new animal model: The NON mouse
Edited by N. Sakamoto, N. Hotta and K. Uchida

Contents

Introduction

The NON mouse is an inbred strain established from the outbred ICR mouse as a control strain of the non-obese diabetic (NOD) mouse by Makino et al. [1, 2]. It is thought to be hereditarily independent from the NOD mouse. In the course of establishing the NON mouse, they were found to have unique renal lesions such as PAS, lipid and IgM positive glomerular intracapillary deposits, dilatation of the straight portion of renal proximal tubules, perivascular lymphocyte infiltrations and hydronephrosis. We present here the characteristics of the glomerular intracapillary deposits and factors influencing their development.

Incidence of glomerular intracapillary deposits

PAS-stained renal sections from 84 males and 99 females at 5 to 63 weeks of age were observed. As shown in Fig. 1, the onset of the glomerular intracapillary deposits was at around 15 weeks of age, and being a little earlier in the females than the males. The deposits seemed to increase with age and the females were more susceptible, though the difference was not statistically significant.

Morphological characteristics of the glomerular intracapillary deposits

Light microscopic observations

The deposits were located in the glomerular capillary lumen as acellular structureless plugs showing slightly eosinophilic, fibrin-like appearances on H–E staining (Fig. 2). The affected glomeruli were randomly localized throughout the cortex and the capillaries were dilated by the plugs. There were no remarkable changes thought to be associated with the plug formations in the glomeruli, interstitium and tubules even in cases having many large deposits. The deposits were strongly positive for PAS (Fig. 3), Sudan III, and Sudan black B, but negative for PTAH, Congo red and periodic acid-silver methenamine stainings (Fig. 4). The sunanophility was markedly

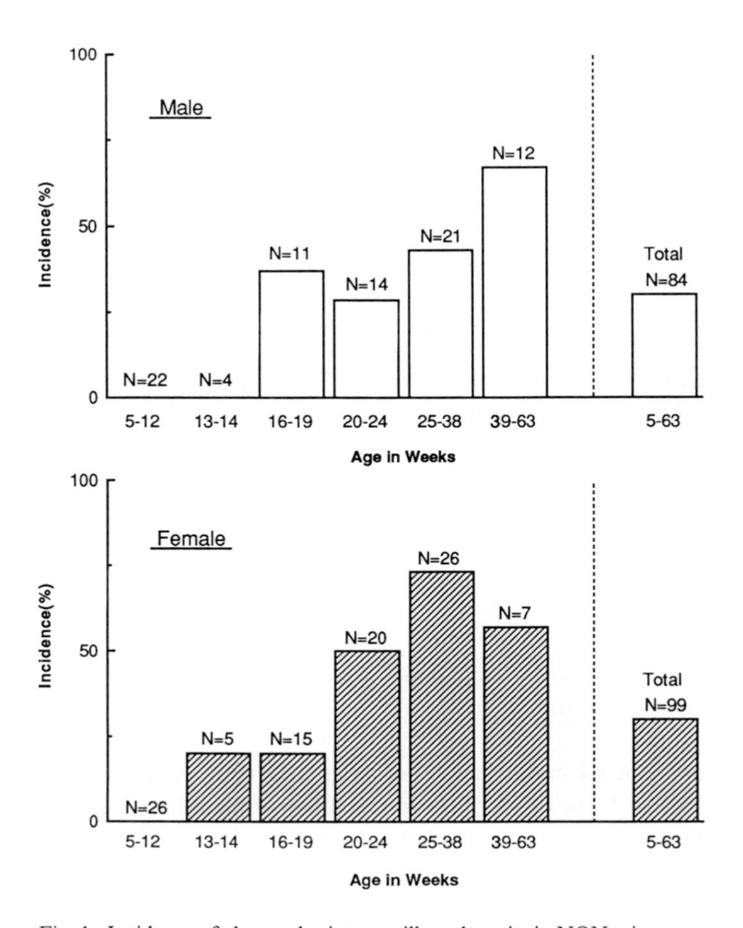

Fig. 1. Incidence of glomerular intracapillary deposits in NON mice.

weakened in the paraffin sections and lost after extraction with chloroform, and PAS positivity disappeared after digestion with trypsin. From these results and the electron microscopic findings cited later, the deposited materials seemed to have a glycoprotein nature incorporating lipids.

There were no remarkable histopathological changes in the organs including the liver, heart, pancreas, spleen, submaxillary gland, adrenal, thyroid, testis, ovary, thymus and lung.

Electron microscopic observations

Transmission electron microscopic observations revealed that the deposits consisted

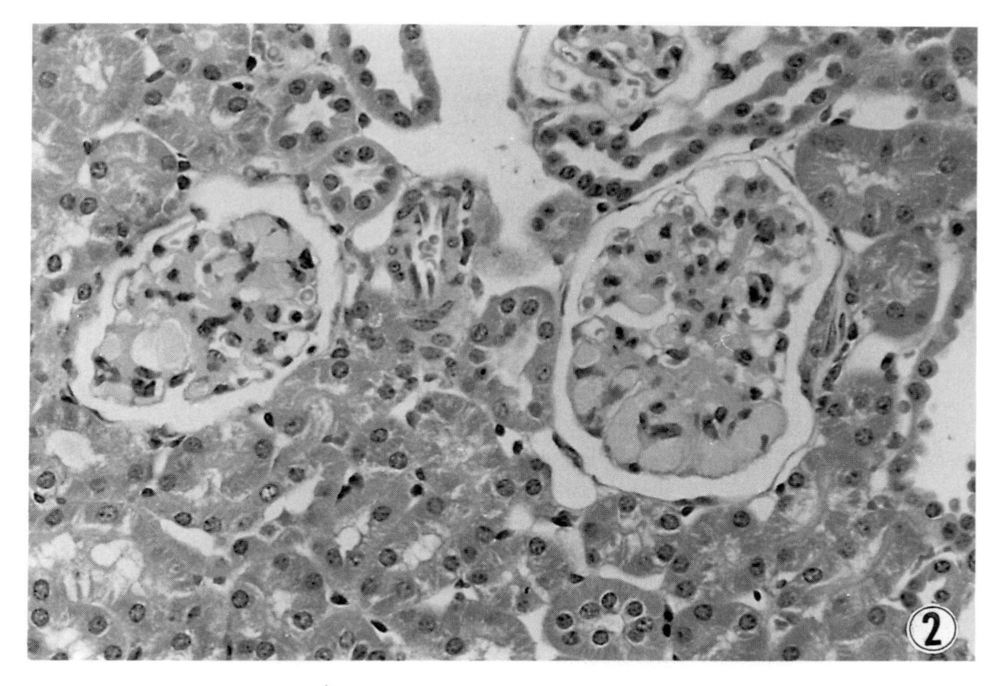

Fig. 2. Kidney from a 16-week-old NON mouse. Weakly stained structureless materials are plugging in the capillary lumens. H-E, original magnification, × 230.

of fine lipid-like droplets less than 300 nm in diameter and fine electron-dense granules. These components sometimes showed heterogeneous or lamellar arrangements and in some areas, the lipid-like droplets were absent. The deposits were located mainly in the glomerular capillary lumen, partially or completely occluding it (Fig. 5). Small amounts of the deposits appeared in the mesangial area, subendothelial zone and the basement membrane near the mesangial stalks (Fig. 6). Other constituents of the glomerulus showed no noticeable alterations, although an increase of mesangial matrix and occasional fusion of foot processes of the epithelial cells were occasionally seen. Scanning electron microscopy revealed that the deposits to be granular-surfaced masses with irregular crests, and partial contact to the endothelial surface (Fig. 7).

Immunohistochemistry

On immunohistochemical staining of the deposits in paraffin or frozen sections, the deposits were strongly positive for IgM (Fig. 8), weakly positive for IgG, very weakly positive for IgA and C3, and negative for fibrinogen. The colloid-gold technique for

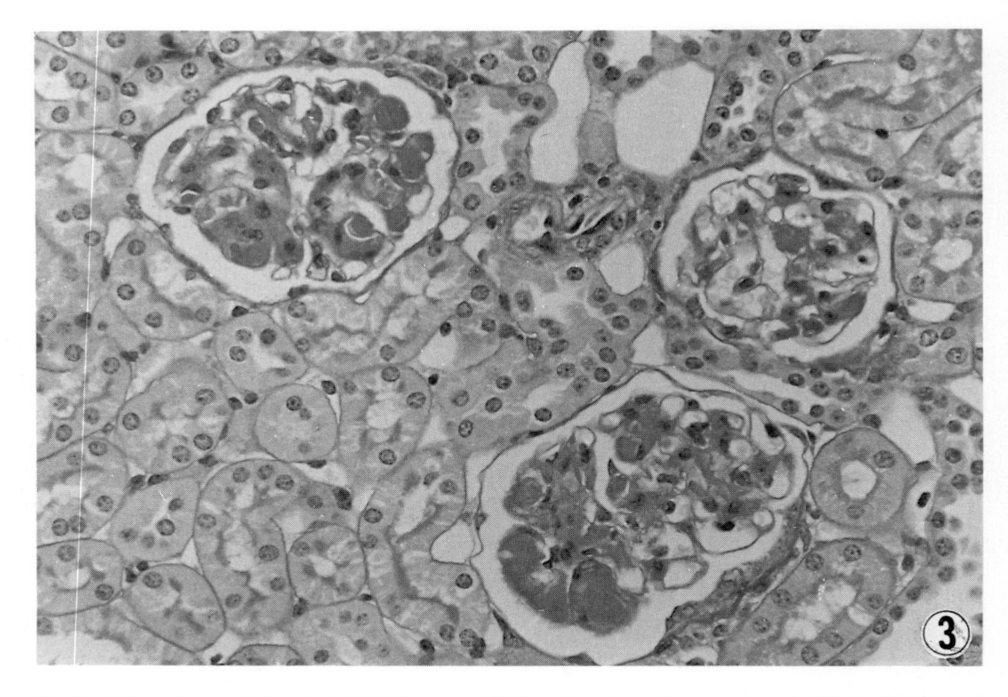

Fig. 3. Kidney from a 16-week-old NON mouse. PAS positive deposits can be seen in the glomerular capillary lumens. PAS, original magnification, × 240.

IgM for electron microscopy demonstrated localization of the gold particles in the capillary lumen and mesangial area in accordance with the findings in the uranium-stained specimens. Whether the IgM and PAS positive deposits were produced in the glomerulus or entrapped blood-borne substances by the specialized glomerular structures remains to be clarified. The glomerular changes in the NON mouse resembled Waldenstrom's macroglobulinemia and cryoglobulinemia in humans in the aspects of PAS positive deposits in the capillary lumen and also resembled the former in the aspect of IgM positive deposits [3]. As the mechanism of the intracapillary deposit formation, Gen [4] stressed increased levels of macromolecular substances in circulation and mesangium overload, and Izuhara et al. [5] reported that the NON mouse may be a useful model for lipoprotein glomerulopathy in humans [6, 7] because of the morphological resemblance of the glomerular changes observed by light and electron microscopy. Recently, Watanabe et al. [15] introduced the NON mouse as a model to investigate lipid disorders in renal diseases.

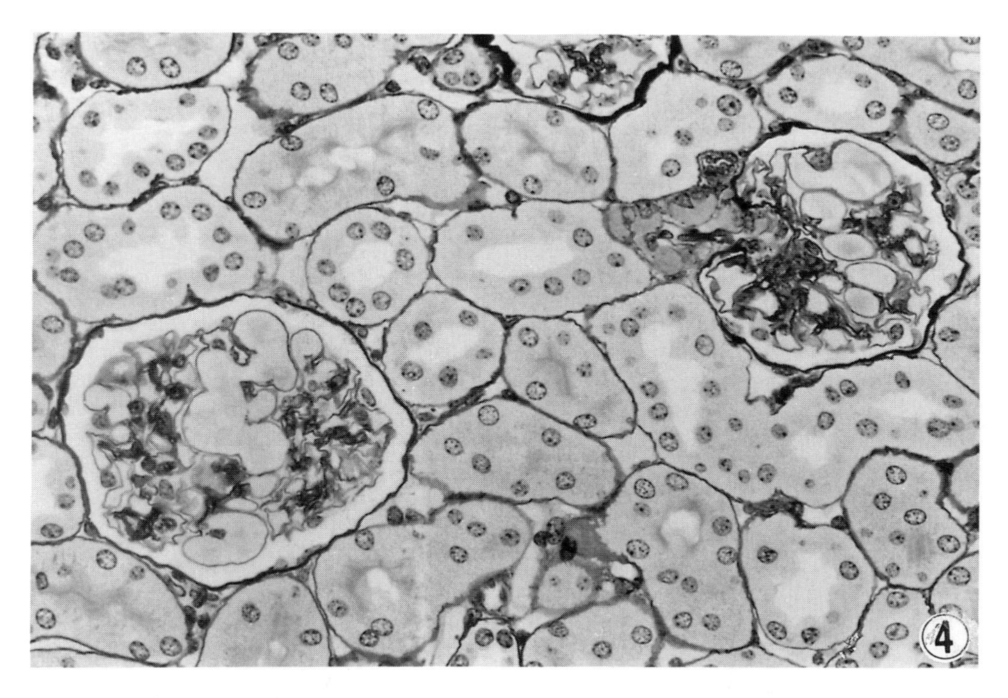

Fig. 4. Kidney from a 16-week-old NON mouse. Deposits that did not stain with PASM were located in the dilated capillary lumens. Periodic acid-silver methenamine, original magnification, × 262.

Renal lesions other than glomerular intracapillary deposits

Other renal lesions included perivascular infiltrations of lymphocytes in both males and females, dilatation of the straight portion of the renal proximal tubules with intracytoplasmic PAS positive droplets of various sizes in the lining epithelial cells, with some protruding into the tubular lumen in the females, and hydronephrosis in the right kidney in the males and females (Table I). Most of the infiltrated lymphocytes in the kidney were Lyt-1 and L3T4 positive, and Lyt-2 positive cells were very rare in immunohistochemical observations, as in the case of insulitis in the NOD mouse [8, 9], suggesting that the lymphocytes mainly consist of helper T cells and may play a role in the pathogenesis of the glomerular lesions if the immune mechanism is involved. Renal tubular changes have been reported by Hadano et al. who suggested that the autoimmune mechanism plays some role in the pathogenesis [10].

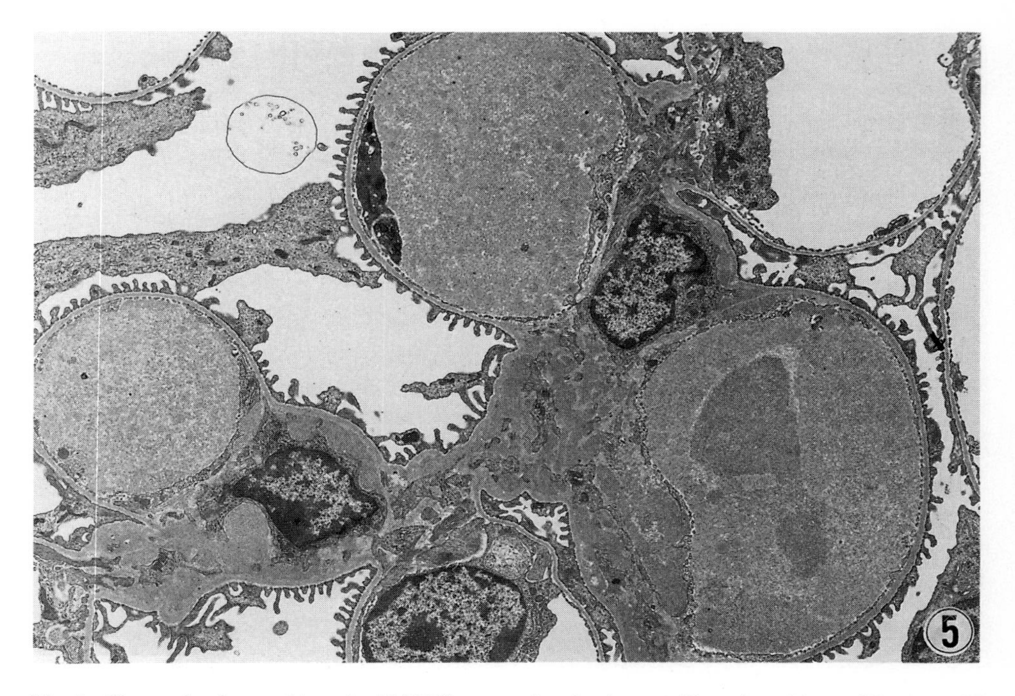

Fig. 5. Glomerulus from a 24-week-old NON mouse showing intracapillary depositions of fine granular materials. The right-most capillary contains deposits with heterogeneous electron densities. Original magnification, × 5800.

Correlations between the glomerular intracapillary deposits and other clinical determinations

NON mice 16 to 26 weeks old were subjected to various clinical laboratory examinations and divided into two groups, those with and without deposits after histopathological observations of the glomeruli. Comparison of the two groups showed no correlations as to growth rates, blood pressure, renal weights, and hematological values, such as erythrocyte and leukocyte counts, differential counts of leukocytes, and plasma levels of urea nitrogen, creatinine, albumin, total protein, total cholesterol and glucose. The values for the platelet count, plasma fibrinogen, urinary fibrinogen degradation products and serum fibrinogen degradation products in the NON mice with deposits were comparable to those in age-matched NON mice without deposits. Urinary protein and sugar were negative for all the mice tested with test paper. Alpha 1 and alpha 2 fractions of the serum protein in the males were 1.5 to 2 times higher than those of the females, but these values did not differ from those in young mice

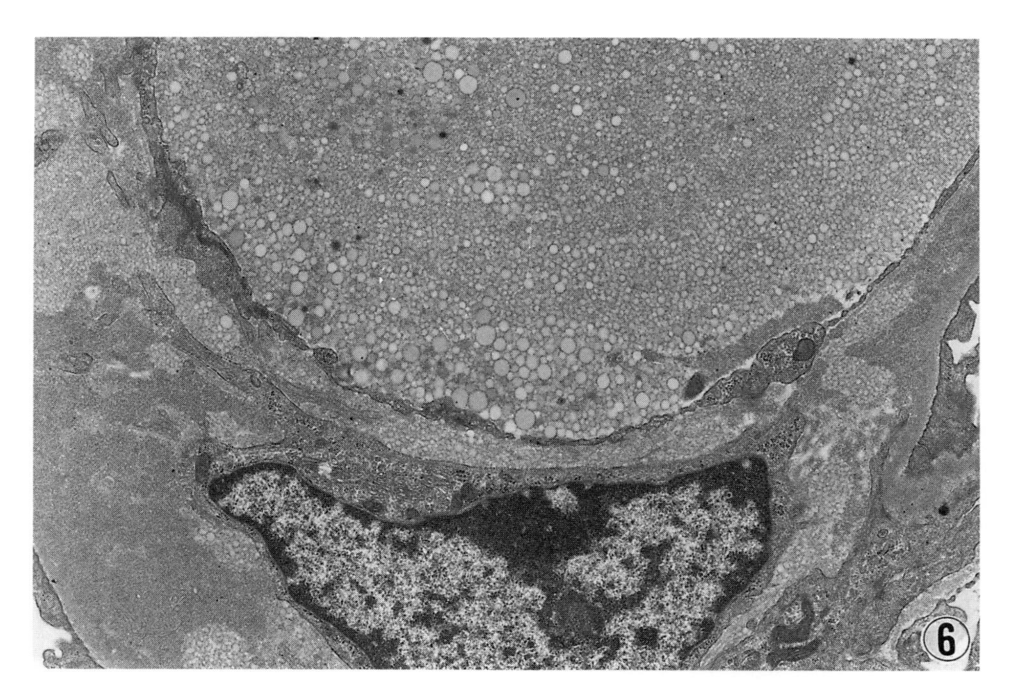

Fig. 6. Glomerulus from a 24-week-old NON mouse. The deposit consisted of fine lipid-like granules of various sizes in the capillary lumen, subendothelial zone and the mesangial area. Original magnification, × 12500.

6 weeks old, the age at which the deposits would not have been formed. The gamma globulin fraction was higher in the aged mice than that of the young mice, but the values were not associated with the deposit formations.

Factors influencing the incidence of glomerular intracapillary deposits and other renal lesions

As mentioned above, the factors thought to influence the deposit formations were aging and sex hormones from sex difference in the incidence but they were not significant. To confirm the effects of the sex hormones and other factors, including unilateral nephrectomy which forces renal overload, and food restriction, which may lessen the renal load, the following experiments were conducted with female NON mice starting when they were 6 weeks old and lasting for 4 months. The mice were divided into eight groups: (1) no treatment, (2) restricted feeding at 70% of group 1, (3) ovariectomy, (4) ovariectomy plus 20 mg/kg testosterone propionate (TP), s.c. × 3/

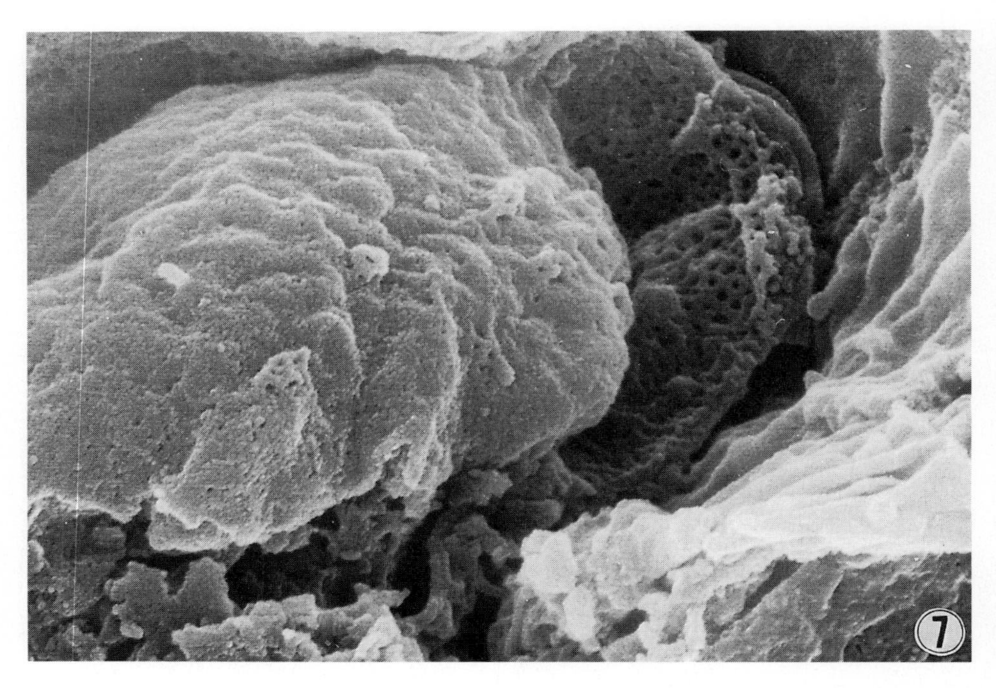

Fig. 7. Glomerulus from a 24-week-old NON mouse. Scanning electron microscopy showing a deposit with a crested granular surface with partial contact to endothelial cell surface. Original magnification, × 35300.

week, (5) ovariectomy plus 400 μg/kg estradiol-17 beta (EST), s.c. × 3/week, (6) unilateral nephrectomy, (7) unilateral nephrectomy plus 20 mg/kg TP, s.c. × 3/week, and (8) unilateral nephrectomy plus 400 μg/kg EST, s.c. × 3/week. Glomerular intracapillary deposits were found in 50 to 80% of all except group 2, with no statistical difference among the groups (Table II). Serum IgM levels determined by an immunodiffusion method were 36.2 $\pm$ 6.4 and 17.0 $\pm$ 4.2 mg/dl in the highest incidence group, group 1, and the lowest incidence group, group 2, respectively. Thus, although the mice in group 1 showed levels twice those of group 2 having no deposits, individual serum IgM levels in the former group did not show correlation with the presence or absence of the deposits or the severity of the deposit formations. Although which dietary factors are effective for inhibition of the renal lesions is unknown, this finding is noticeable because reduction of dietary intake is prophylactic in relation to glomerular lesions involving the immune mechanism in many strains of mice including NZB [11], NZB × NZW [12] or KK [13] mice.

Dilatation of the straight portions of the proximal tubules, which is a characteristic of the females, were found in 60, 40, 90, 90 and 100% animals in groups 1, 3, 5, 6,

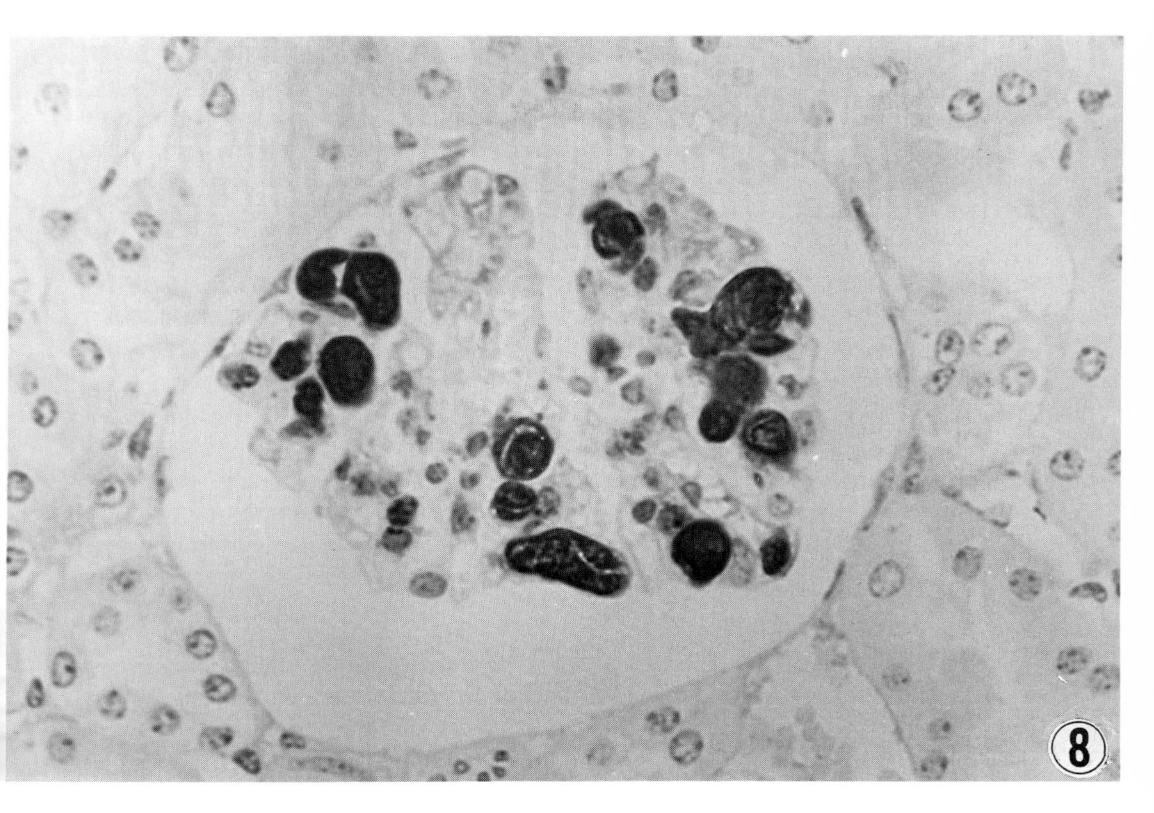

Fig. 8. Glomerulus from a 20-week-old NON mouse showing IgM positivity of the deposits. ABC method, original magnification, × 780.

and 8, respectively, without significant differences from group 1. Groups 2, 4 and 7 showed markedly reduced incidences of the tubular lesions at 0, 0 and 10%, respectively. These findings suggest androgen as a factor suppressing the tubular lesion, which is interesting as androgen suppresses the appearance of diabetes in the NOD mouse [14]. There were no significant differences in the renal perivascular lymphocyte infiltrations among the groups.

The pronounced effect of restricted feeding on the deposit formations led us to examine whether this treatment could remove already present deposits. NON mice 20 to 23 weeks old were unilaterally nephrectomized and the glomeruli having deposits were counted in the excised right kidneys. After the operation, the mice were kept under free feeding or restricted feeding for four months, and then the glomeruli having deposits in the left kidneys were counted. The results revealed reduced incidence in four of five mice in the restricted feeding group in comparison to that in the right

TABLE I
Pathological findings of NON mice[a]

Pathological findings	Mice with deposits		Mice without deposits		Total	
	Male	Female	Male	Female	Male	Female
No. of mice	25	40	32	28	57	68
Lymphocyte infiltration						
Kidney	14(56.0)[c]	15(37.5)	13(40.7)	15(53.6)	27(47.4)	30(44.1)
Liver	4(16.0)	7(14.5)	5(15.6)	7(25.0)	9(15.8)	14(20.6)
Submaxillary gland	0	2(5.0)	0	0	0	2(2.9)
Pancreas (exocrine portion)	3(12.0)	5(12.5)	2(6.2)	3(10.7)	5(8.8)	8(11.8)
Renal tubular dilatation[b]	0	16(40.0)	0	9(32.1)	0	25(36.8)
Renal glomerular sclerosis	0	1(2.5)	0	0	0	1(1.5)
Hydronephrosis (right)	5(20.0)	2(5.0)	5(15.6)	3(10.7)	10(17.5)	5(7.4)

[a]16 or more weeks old.
[b]Straight portion of proximal tubule.
[c]Figures in parenthesis indicate % values.

kidneys four months earlier, and no difference for one mouse. In the free feeding group, seven of twelve mice showed increased incidence of the deposits and no difference for the remaining five mice. These findings suggest that there may be a reversible period for glomerular lesions and that dietary factors may manipulate the incidence of the deposits.

Conclusion

Characteristic pathological changes of the NON mouse include PAS, lipid and IgM positive deposits which were seen only in the renal glomerular capillary lumen. These lesions occur from around 15 weeks of age and tend to increase with aging, while restricted feeding can markedly reduce this incidence. Other organ changes include dilatation of the straight portion of the renal tubules, which is seen only in the females, and lymphocyte infiltrations, which appear mainly in the kidney and liver. It is interesting that there were no notable changes thought to be related to the deposit formation in histopathological observations of the organs and clinical laboratory determinations, even in the mice with severe deposits. The pathogenesis of the glomerular lesion and usefulness of this disease model remain to be clarified.

TABLE II

Influence of various treatments on the incidence of glomerular intracapillary deposits in NON mice

Treatment (4 months)	No. of mice	Final body wt	Renal wt	No. of mice with deposits	% incidence of glomerulus with desposits				
		(mean ± SD, g)			0	<5	5–25	25–50	50<
Nontreatment	10	40.6 ± 4.6	0.40 ± 0.07	8(80)[a]	2[b]	0	4	2	2
Food restriction	10	24.9 ± 1.7[e]	0.30 ± 0.07[e]	0(0)[e]	10	0	0	0	0
Ovariectomy	10	41.4 ± 5.5	0.40 ± 0.12	6(60)	4	0	2	2	2
Ovariectomy + TP[c]	10	38.0 ± 2.0	0.46 ± 0.11	5(50)	5	2	2	1	0
Ovariectomy + Est[d]	10	38.9 ± 1.2	0.48 ± 0.09	7(70)	3	1	3	3	0
Unilateral nephrectomy	10	35.2 ± 3.4[e]	0.36 ± 0.06	8(80)	2	2	4	1	1
Unilateral nephrectomy + TP[c]	10	36.7 ± 2.3[e]	0.35 ± 0.03	5(50)	5	0	2	3	0
Unilateral nephrectomy + Est[d]	10	34.9 ± 3.3[e]	0.35 ± 0.04	6(60)	4	2	3	1	0

[a]Figures in parenthesis indicate % values.
[b]Number of mice.
[c]20 mg/kg s.c. testosterone propionate three times a week for 4 months.
[d]400 μg/kg s.c. estradiol-17β three times a week for 4 months.
[e]Significant difference from nontreatment group at $P < 0.05$.

Acknowledgements

The authors wish to thank Dr. M. Harada, Mr. K. Takano and J. Ikeuchi for their valuable advice and technical support in the present study.

References

1 Makino S, Kunimoto K, Muraoka Y, Mizushima Y, Katagiri K, Tochino Y. Breeding of non-obese, diabetic strain of mice. Exp Anim 1980;29:1–13.
2 Makino S, Hayashi Y, Muraoka Y, Tochino Y. Establishment of the nonobese diabetic (NOD) mouse. In: Sakamoto N, Min HK, Baba S, eds. Current topics in clinical and experimental aspects of diabetes melitus. Amsterdam: Elsevier, 1985;25–32.
3 Churg J, Sobin LH. Renal disease – Classification and atlas of glomerular diseases. Tokyo: IGAKU-SHOIN, 1982;240–260.

4 Gen E. Histopathological observations of spontaneous development of glomerular intracapillary deposits in mice – its relation to macromolecular clearance. Wakayama Med Res 1979;21:127–140.
5 Izuhara M, Suzuki N, Kagami H, Yamazaki C, Watanabe Y, Yoshida F, Matsuo S. Histopathological study of NON (nonobese-nondiabetic) strain of mice with lipid deposition in the glomerular capillary lumen. J Clin Electron Microscopy 1989;22:5–6.
6 Saito T, Sato H, Kudo K, Oikawa S, Shibata T, Hara Y, Yoshinaga K, Sakaguchi H. Lipoprotein Glomerulopathy: Glomerular lipoprotein thrombi in a patient with hyperlipoproteinemia. Am J Kidney Dis 1989;13:148–153.
7 Watanabe Y, Ozaki I, Yoshida F, Fukatsu A, Itoh Y, Matsuo S, Sakamoto N. A case of nephrotic syndrome with glomerular lipoprotein deposition with capillary ballooning and mesangiolysis. Nephron 1989;51:265–270.
8 Hanafusa T, Nonaka K, Tarui S. Immune pathogenesis of the NOD mouse – An overview. In: Tarui S, Tochino Y, Nonaka K, eds. Insulitis and type I diabetes, lessons from the NOD mouse. Tokyo: Academic Press, 1986;75–82.
9 Miyazaki A, Nonaka K, Tarui S. Lymphocytic subsets in islets and spleen of the NOD mouse. In: Tarui S, Tochino Y, Nonaka K, eds. Insulitis and type I diabetes, Lessons from the NOD mouse. Tokyo: Academic Press, 1986;83–90.
10 Hadano H, Suzuki S, Tanigawa K, Ago A. Cell infiltration in various organs and dilatation of the urinary tubule in NON mice. Exp Anim 1988;37:479–48.
11 Fernandes G, Yunis EJ, Good RA. Influence of protein restriction on immune functions in NZB mice. J Immunol 1976;16:782–790.
12 Friend PS, Fernandes G, Good RA, Michael AF, Yunis EJ. Dietary restrictions early and late effects on the nephropathy of the NZB × NZW mouse. Lab Invest 1978;38:629–632.
13 Emoto M, Matsutani H, Shirakawa H, Kimura S, Sakaue K, Kadoya Y, Miyamura K, Miyajima M. Natural history of glomerular lesion in KK mice and its regression by feeding with limited diet. Wakayama Med Rep 1982;25:7–18.
14 Makino S, Kunimoto K, Muraoka Y, Katagiri K. Effect of castration on the appearance of diabetes in NOD mouse. Exp Anim 1981;30:137–140.
15 Watanabe Y, Itoh Y, Yoshida F, Koh N, Tamai H, Fukatsu A, Matsuo S, Hotta N, Sakamoto N. Unique glomerular lesion with spontaneous lipid deposition in glomerular capillary lumina in the NON strain of mice. Nephron 1991;58:210–218.

CHAPTER 11

Cell infiltration in various organs and dilatation of the urinary tubule in the NON mouse

KEIICHIRO TANIGAWA[a], SYUSAKU SUZUKI[b], HITOMI SAHATA[c], and YUZURU KATO[a]

[a]Department of Medicine, Shimane Medical University, Izumo, Shimane, Japan
[b]Institute of Experimental Animals, Shimane Medical University, Izumo, Shimane, Japan
[c]Shimane Institute of Health Science, Shimane Medical University, Izumo, Shimane, Japan

Current concepts of a new animal model: The NON mouse
Edited by N. Sakamoto, N. Hotta and K. Uchida

Contents

Introduction

Animal models such as non-obese diabetic (NOD) mouse are very valuable to understand pathogenesis, complication and treatment of insulin-dependent diabetes mellitus (IDDM). Non-obese non-diabetic (NON) mouse was obtained from a subline of the NOD mouse, but NON mouse does not develop diabetes in spite of impaired glucose intolerance [1, 2]. In the present article, some characteristic features of NON mouse are described. Those are cell infiltration in various organs and dilatation of the urinary tubule [3]. Since IDDM in NOD mouse greatly develops in female rather than in male animals [1], it is suggested that sex hormones play an important role in causing IDDM in NOD mouse [4]. In analogy, we tested the hypothesis whether sex hormones are involved in the development of cell infiltration in the characteristic dilatation of the urinary tubule in NON mouse.

Materials and methods

NON mice were obtained from the colony maintained at Shionogi Research Laboratories (Osaka, Japan) and maintained at the Institute of Experimental Animals, Shimane Medical University, under closed condition as previously described in detail [3]. In the present study, we examined the animals at 30, 60, 90, 150, 200, 250 and 300 days of age. As a control, Jcl: ICR mouse at 120 days of age was used. The mice were exsanguinated under light ether anesthesia and tissues were quickly removed. They were fixed in a Bouin's solution. Paraffin-embedded sections were stained by hematoxylin and eosin, Masson-trichrome (M-T) and periodic acid Schiff (PAS). The portion of the kidney was double-fixed in 2.5% glutaraldehyde-1% osmium tetroxide for transmission electron microscopy (TEM). For scanning electron microscopy (SEM), the tissue blocks were fixed in a mixture of 2.5% glutaraldehyde and 5% paraformaldehyde. They were then immersed in 50% dimethylsulfoxide and fractured in liquid nitrogen, and were dried by the critical point drying method.

To evaluate the role of sex hormones on tubular dilatation in the kidney, male mice were orchidectomized at 30 days of age. Those animals were killed at 90 and 150 days of age and the kidney was removed for histological examination as described above.

We also examined the effect of estrogen administration (20 μg, s.c., once a week) on the tubular lesion in a castrated male NON mouse from 60 days to 150 days of age.

Results

Cell infiltration

We found cell infiltration in 6 out of 10 organs examined, which appeared at different periods and various incidences as shown in Table I. Most of the infiltrated cells were lymphocytes and a few plasmacytes and monocytes, as have been extensively studied in the NOD mouse [5]. There were numerous cell infiltrations in the kidney, exorbital lacrimal gland and mandibular gland. The infiltration of lymphocytes was observed in almost all the kidneys of both sexes after 60 days of age. The infiltration took place around the arcuate-artery and -vein near the papillary duct, which became more manifest with aging. At 250 days of age, more cell infiltration was found around other vessels rather than the arcuate-artery and -vein (Fig. 1). Cytoplasma of tubule cell was resulted in destruction by cell infiltration, which was filled in the lumen. There was a tendency that the grade of cell infiltration in the kidney from female animals was greater than that from male animals.

In the submaxillary gland, cell infiltration was observed at 60 days of age and markedly increased after 200 days of age. Cell infiltration in the exorbital lacrimal gland was more prominent in female mouse than in male mouse. In contrast to the NOD mouse, only modest cell infiltration was observed in Langerhans' islets of the pancreas at 60 days of age. Similarly, there is some cell infiltration in the thyroid gland and the parotid gland. We did not observe any cell infiltrations in the ovary, testis, adrenal gland and sublingual gland.

Tubular dilatation

The most peculiar finding observed in the NON mouse was the dilatation of the tubules after 60 days of age, which became strongly apparent with aging (Fig. 1). On the other hand, we did not observe any dilatation of the tubule from male NON mice. Therefore, we describe here the finding obtained in female NON mice.

Slight dilatation of the tubules took place from the border between cortex and medulla to the external medulla at 60 days of age and such dilatation of the proximal tubules became apparent at 90 days of age. Figure 2 illustrates numerous acidophil bodies in cytoplasma and lumen of the dilated proximal tubule that were intensively stained with PAS and showed blue color with Masson-trichrome staining. Figures 3 and 4 reveal the luminal surface of the dilated proximal tubule observed by electron microscopy. There are numerous microvilli extending into the lumen. An apocrine

TABLE I
Incidence of cell infiltration (%) in various organs. The number of (+) signs indicate the quantitative degree ranging from weak (+) to strong (+ + +). There was no cell infiltration in the ovary, testis, adrenal gland, and sublingual gland

Day	30		60		90		150		200		250		300	
Organ / Sex	M	F	M	F	M	F	M	F	M	F	M	F	M	F
Kidney	0	0	100	60	100	80	100	100	100	100	100	100	100	100
Exorbital Lacrimal g.	0	0	(+) 40	(+) 20	(+) 80	(+−++) 60	(+−++) 100	(++−+++) 75	(++) 62	(+++) 100	(++−+++) 66	(+++) 66	(++−+++) 100	(+++) 80
Mandibular g.	0	0	(+) 20	(+) 0	(+) 60	(+−++) 80	(+−++) 100	(+−++) 100	(++) 62	(++) 100	(++) 100	(++) 100	(++) 100	(++) 100
Pancreas	0	0	(+) 20	0	(+) 0	(+) 20	(+) 0	(+−++) 50	(+−++) 12	(++) 0	(+−++) 16	(++) 66	(++) 0	(++) 20
Thyroid g.	0	0	(+) 0	0	0	(+) 20	0	(+) 25	(+) 0	0	(+) 0	(+) 0	0	(+) 0
Parotid g.	0	0	0	0	0	(+) 20	0	(+) 0	0	0	0	16	0	0

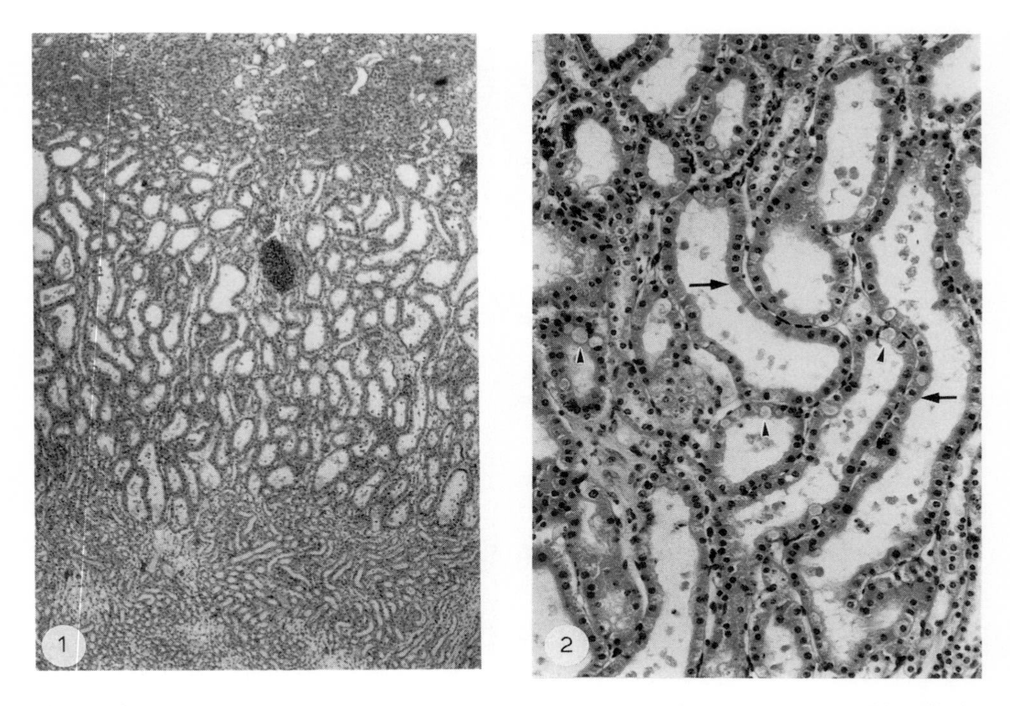

Fig. 1. Light micrograph of the kidney in NON mouse (300-day-old female). Note the striking dilation of the urinary tubule and the infiltrating cell masses. H.E. stain 4 × 3.3.

Fig. 2. Light micrograph of the dilated proximal tubule in a NON mouse (300-day-old female). Numerous acidophil bodies (arrowheads) are observed in the cytoplasm. Brush borders (arrows) are present in the apical surfaces of dilated tubuli. H.E. stain 20 × 2.5.

process with various structures and containing different particles projects into the lumen. Smaller and larger spherical protrusions are best shown by scanning electron microscopy (Fig. 4). These apocrine process are not present in the kidney from ICR mouse and male NON mouse.

Further dilatation of the tubule was noted, and more PAS-positive bodies in cytoplasma and intra-lumen were found at 90 days of age. In addition, slight degeneration was noted in the dilated tubular cells. At 250 days of age, tubular dilatation was further striking in the external medulla. The dilatation was also noticeable in the tubule of the cortex and the internal medulla and associated with a lot of PAS-positive bodies. Degeneration of cytoplasma was progressive and resulted in eosinophilic substances in the lumen, presumably from partial detachment of degenerated cells. The PAS-positive cast was seen in the tubular lumen from internal to papilla of the medulla. These changes were more striking and severe, including vacuolation of cytoplasma and disappearance of microvilli in a 300-day-old mouse.

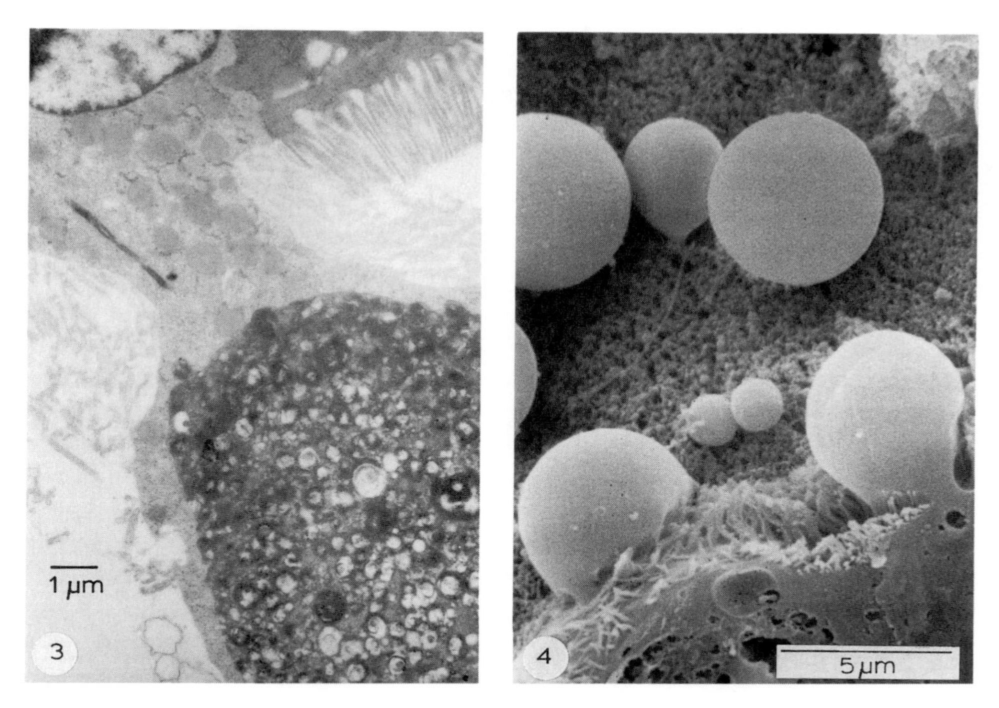

Fig. 3. TEM. Luminal surface of the dilated proximal tubule in a NON mouse (300-day-old female). An apocrine process projects into the lumen. This process contains complex structures. Numerous microvilli extend into the lumen.

Fig. 4. SEM. Luminal surface of the dilated proximal tubule in a NON mouse (300-day-old female). Note the spherical protrusions and numerous microvilli extending into the lumen.

Effect on tubular dilatation of castration in male NON mice

At 90 days of age, a slight dilatation of the tubule and the PAS-positive bodies was noticed in the kidney from castrated male mice. Further dilatation and more PAS-positive bodies were demonstrated at 150 days of age. Of special interest was the fact that administration of estradiol induced prominent dilatation of the tubule and greater number of PAS-positive bodies in the kidney from castrated male NON mice (Figs. 5 and 6).

When a male ICR mouse as a control was castrated, there was no histological change in the kidney at 120 days of age. Slight dilatation of the tubule and a few PAS-positive bodies were seen at 150 days of age, which were almost of a similar grade as observed in castrated NON mouse at 90 days of age.

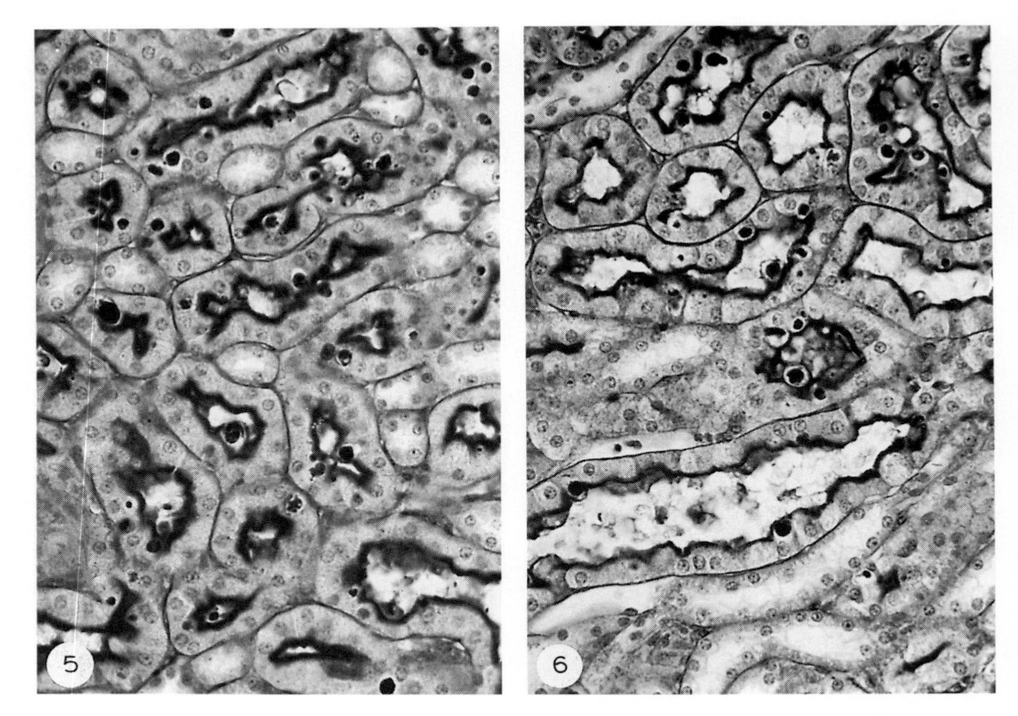

Fig. 5. Light micrograph of the proximal tubule in a castrated NON mouse (150-day-old male). The dilatation of the urinary tubules and the PAS-positive bodies are observed. PAS. 20 × 3.3.

Fig. 6. Light micrograph of the proximal tubule in a castrated NON mouse treated with estradiol (150-day-old male). The prominent dilatation of the urinary tubule and PAS-positive bodies are noticed. PAS. 20 × 3.3.

Discussion

There are controversial reports regarding glucose intolerance in NON mice. We tested i.p. glucose tolerance test and found that a male NON rather than a female NON mouse was slightly glucose intolerant [6]. However, we have never observed glycosuria until an age of 365 days in these animals. Thus, the finding that female NON mice showed a peculiar lesion such as dilatation of the proximal tubule with many PAS-positive bodies is not directly associated with glucose intolerance. In this context, we further tested the possibility whether hyperglycemia itself could modulate tubular lesions [6]. 60-day-old female NON mice received i.p. injection of streptozotocin (120 mg/kg). They were killed at 220 days of age. We did not find any cellular proliferation and hypertrophy of the kidney. STZ-induced diabetes did not affect dilatation of the proximal tubule, but resulted in hydronephrosis. Thus, these findings indicate that hyperglycemia itself is not involved in kidney lesion observed in female NON mice. Rather, sex hormones and genetic factors could be involved.

The pathogenesis of Type I diabetes has been extensively studied using NOD mouse. However, there are few reports regarding the characterization of the NON mouse. The previous reports [1, 2] and our observations [3, 6] suggest that the NON mouse is not a suitable animal model for the study of Type II diabetes. Although the lymphocyte infiltration is found in pancreatic islets, it is hard to observe the destruction of pancreatic β cells, in contrast to the findings obtained in female NOD mice.

The possibility that autoimmunity is attended with tubular dilatation in female NON mice remains unclear, since cell infiltration is found in both sexes. Makino et al. [4] revealed that oophorectomy in female NOD mice prevents the occurrence of Type I diabetes. Thus, it is tempting to speculate that sex hormones play a crucial role in developing diabetes in NOD mice and in tubular dilatation in NON mice. Our observations that orchidectomy induce tubular dilation in male NON mice indicate that the male sex hormones prevent tubular dilatation. In addition, the finding that estradiol administration causes further dilatation of tubules and numerous PAS-positive bodies, is inevitable evidence indicating that the female sex hormones play a regulatory role in tubular dilatation.

Acknowledgement

We are grateful to Mrs. A. Kawakami for typing the manuscript.

References

1 Makino S, Kunimoto K, Muraoka Y, Mizushima Y, Katagiri K, Tochino Y. Bleeding of a non-obese, diabetic strain mice. Exp Anim 1980;29:1–13.
2 Tochino Y. Discovery and bleeding of the NOD mouse. In: Tarui S, Tochino Y, Nonaka K, eds. Insulitis and type I diabetes. Lessons from the NOD mouse. Tokyo: Academic Press Japan, 1986;3–10.
3 Hadano H, Suzuki S, Tanigawa K, Ago A. Cell infiltration in various organ and dilatation of the urinary tubule in NON mice. Exp Anim 1988;37:479–483.
4 Makino S, Kunimoto K, Muraoka Y, Katagiri K. Effect of castration on the appearance of diabetes in NOD mouse. Exp Anim 1981;30:137–140.
5 Asamoto H, Oish M, Akazawa Y, Tochino Y. Histologic and immunologic changes in the thymus and other organs in NOD mice. In: Tarui S, Tochino Y, Nonaka K, eds. Insulitis and type I diabetes. Lessons from the NOD mouse. Tokyo: Academic Press Japan, 1986;61–74.
6 Tanigawa K, Kirihara Y, Kato Y, Suzuki S. Effect of diabetes mellitus on kidney lesion in NON mouse. In: Goto Y, ed. Diabetic animals Vol 3, Medical Journal 1989; 240–245 (in Japanese).
7 Takai I, Maruyama T, Taniyama M, Kataoka K. Humoral immunity in the NOD mouse. In: Tarui S, Tochino Y, Nonaka K, eds. Insulitis and type I diabetes. Lessons from the NOD mouse. Tokyo: Academic Press Japan, 1986;101–110.

CHAPTER 12

Immunopathological observations of nephropathy and characterization of infiltrated lymphocytes

YUZO WATANABE, YASUHIKO ITOH, DAIJO MIZUMOTO, FUTOSHI YOSHIDA, NAOKI KOH, NIGISHI HOTTA and NOBUO SAKAMOTO

Third Department of Internal Medicine, Nagoya University School of Medicine, Tsuruma, Nagoya, Japan

Current concepts of a new animal model: The NON mouse
Edited by N. Sakamoto, N. Hotta and K. Uchida

Contents

Introduction

Diabetic nephropathy is a serious secondary complication of diabetic patients. Though various diabetic animal models have been reported already, no model showed the typical glomerular lesion which resembled human diabetic nephropathy [1, 2]. Since the progression of diabetic nephropathy is insidious, the establishment of diabetic animal models seems to be a useful tool to investigate the pathogenesis of diabetic nephropathy. Non-obese non-diabetic (NON) mice were developed together with a famous diabetic strain of mice named non-obese diabetic (NOD) by Makino [3]. Contrary to the naming of NON, this strain of mice was reported to show impaired glucose tolerance (IGT) [4, 5]. Therefore, we carried out a histological analysis on this strain of mice to investigate whether this mice had the characteristic diabetic lesion or not. We found a variety of morphological abnormalities in this strain of mice. These lesions consisted of glomerular lesion, tubular lesion, and lymphoid follicle-like structures around renal arterioles. Among them, glomerular lesion was a peculiar one and resembled diabetic nodular and/or exudative lesions. Meanwhile, the lymphoid follicle-like structures were commonly observed in aged NON mice, which resembled the lymphocytic infiltrates presented around pancreatic islets in NOD mice. However, the essential nature of the glomerular lesion was somewhat different from that of diabetic nephropathy and was characterized by massive lipid deposition in glomerular capillary lumina. This lesion was observed irrespective of the existence of IGT. Therefore, we now suppose that this glomerular lesion does not have any relation with IGT.

Thus, we conducted this study to clarify the histopathological nature and frequency of these lesions paying attention to the relationship between renal morphological abnormalities and IGT. In addition, we tried to explore the immunological abnormalities of NON mice by measuring serum concentration of immunoglobulins as well as characterization of infiltrated lymphocytes, because lymphoid follicle-like structures were one of the characteristic morphological abnormalities which were found in NON mice.

Materials and methods

The NON strain of mice was a generous gift from Aburahi Laboratories, Shionogi & Co., Ltd., Japan. These mice were maintained at the Animal Breeding Center of our institute with free access to water and standard mouse chow. In total, 107 NON mice (male 57, female 50, 3–12 months old) were submitted to this study. As a control strain of NON mice, we purchased 3- and 6-month-old outbred ICR strain of mice (male 10, female 10) from Clea Japan Inc. This ICR strain is the maternal strain from which the NON mouse was developed.

Glucose tolerance test

Glucose tolerance test was done on a number of mice by intraperitoneal glucose administration with a dose of 2 g/kg body weight after a minimum of 16 hour of food deprivation. Serial samples (i.e. before glucose load, 1 and 2 hour after loading) were drawn from mice orbita under light anesthesia and examined immediately on a glucose analyzer (Nikkaki Co., Tokyo, Japan) using the glucose oxidase method.

Histopathological study of the kidney

Renal tissues, which were obtained at sacrifice, were submitted to histopathological study in all mice of both the NON and ICR strain. Renal tissue was divided into three portions and processed for light (LM), immunofluorescence (IF), and electron microscopy (EM). IF and EM were done in some of them. The specimens of ICR strain of mice were used as control.

Specimens for LM were fixed in 10% buffered formate and embedded in paraffin. Hematoxylin and eosin (HE), periodic acid Schiff (PAS), and periodic acid methenamine silver (PAM) staining were done. To investigate the lipids deposition, snap-frozen tissues unfixed or fixed in Baker's fixative were stained by the methods of Sudan III, Nile blue sulfate. Sudan black B, Schultz stainings with Digitonin, and salt extraction [6]. For IF, the tissue was snap-frozen in liquid nitrogen and cryostat sections were stained by the direct IF technique. The direct IF was done with FITC-labelled goat anti-mouse IgG, A, and M, and C3 (Cappel Laboratories, Cochranville, PA, USA). EM tissues were fixed in 2.5% glutaraldehyde, postfixed with 1% osmium tetroxide, and embedded in EPON 812. Ultrathin sections stained with uranyl acetate and lead citrate were examined in a JEOL-100CX microscope (Tokyo, Japan).

To clarify the cell type of lymphocytes in lymphoid follicle-like structures, immunohistochemical staining was done according to the method of Nakane and Pierce [7]. Rat monoclonal antibodies to mouse Ly-1, Ly-2, and L3T4 were used for staining (Beckton–Dickinson Immunodiagnostics, Mountain View, CA, USA). The positive stainings against the antibodies used were considered as Ly-1: pan T cell, Ly-2; suppressor T cell, and L3T4: helper T cell.

Histopathological study of other organs

A variety of organs such as pancreas, brain, lung, spleen, liver, lymph nodes, thyroid gland, and aorta were obtained at sacrifice in some of the mice. These tissues were processed for LM specimens as indicated above.

Serum immunoglobulin levels

To clarify the immunological disarrangement of NON mice, serum Ig concentrations in mice sera were measured by enzyme-linked immunoadsorbent assay (ELISA). ELISA was done with flat-bottom polyvinylchloride plates (Immulon II, Dynatech Laboratories Inc., Chantilly, VA, USA). The principle was solid-phase sandwich enzyme immunoassay.

Subpopulation of lymphocytes in spleen

To characterize the subpopulation of T-lymphocytes clusters, the lymphocytes in spleen were collected and crossreacted with rat monoclonal antibodies against mouse Ly-1, Ly-2, and L3T4. Thereafter the cells were differentiated by flowcytometry with EPICS-PROFILE (Coulter, Hialeah, FL, USA). The positive staining cells against those antibodies were classified as mentioned above.

Results

Serial body weight changes

The serial changes of body weight are shown in Fig. 1. NON mice showed continuous weight gain up to 12 months of age in both sexes. There was less body weight gain in female than male in NON mice. However, there was no significant difference of body weight between NON and ICR mice. This result confirms that the NON strain is not an obesity animal model.

Glucose tolerance test

Results of GTT which were done in various age groups of mice individually are shown in Table I. High blood glucose level after glucose loading was seen in the NON group, especially in male from 3 months of age onwards. There was a statistically significant difference in blood glucose levels after glucose loading between NON and ICR mice. When we settled the normal upper limit of the blood glucose value as 350 mg/dl, impaired glucose tolerance (IGT) was observed frequently in NON mice.

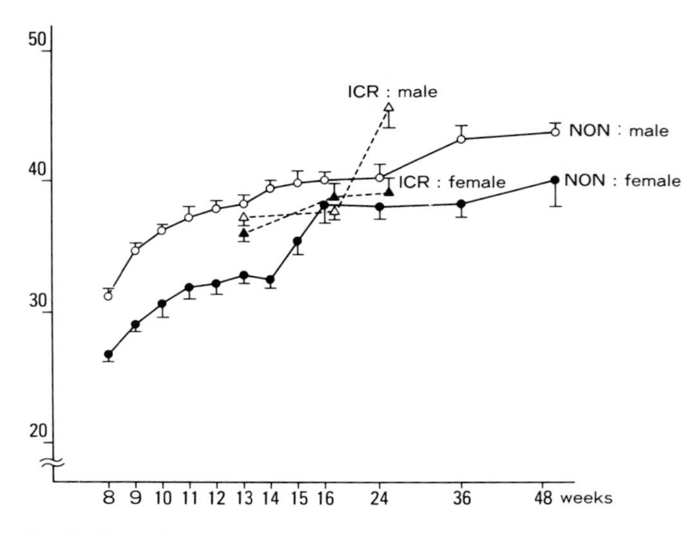

Fig. 1. The serial changes of body weight of NON mice and mice of the ICR strain. The values are expressed as mean ± SE. There was no statistical difference between the two strains of mice.

TABLE I

The mean values of glucose tolerance test in various age classes of the NON and ICR strain of mice. There are statistically significant differences between NON and ICR strain of mice. (Values are expressed as mean ± SE, unit: mg/dl.)

Age (months)	ICR				NON							
					female				male			
	Vor	1 hr	2 hr		Vor	1 hr	2 hr		Vor	1 hr	2 hr	
3	139	232	160	(10)	110	240	176	(5)	140	478[b]	289[b]	(5)
	±5	±10	±7		±6	±10	±10		±8	±73	±58	
4	–	–	–		117	255	199	(8)	135	445	381	(8)
					±8	±14	±9		±14	±61	±60	
6	128	275	181	(10)	110	250	194	(13)	110	387[a]	322[b]	(21)
	±4	±16	±13		±7	±15	±7		±4	±34	±34	
9	–	–	–		112	542	409	(11)	142	613	671	(6)
					±6	±61	±50		±7	±92	±107	
12	–	–	–		118	276	245	(10)	164	457	379	(9)
					±10	±58	±53		±23	±66	±71	

[a] $P<0.05$
[b] $P<0.01$

Meanwhile, IGT was not observed in ICR mice. The overall rate of IGT in NON mice was 49.0% (47/96). There was a sex difference in the frequency of IGT, and its incidence was higher in males than females (male: 65.3%, 32/49; female: 31.9%, 15/47).

Urinalysis

Urinalysis was done using Tes-tape (Ames, Tokyo, Japan). Glucosuria was not found in NON mice even at the postprandial state, though IGT was found in about half of the NON mice. The amount of proteinuria was always higher in male than in female mice, in both NON and ICR mice. The degree was higher in NON mice than ICR mice in both sexes.

Renal morphological changes

Three major renal lesions were found in NON mice, namely glomerular lesion, tubular lesion, and lymphoid follicular-like lesion (lymphoid lesion). The incidence and the degree of these lesions were gradually increased by aging (Fig. 2). These lesions were specific in NON mice and they were never seen in ICR mice.

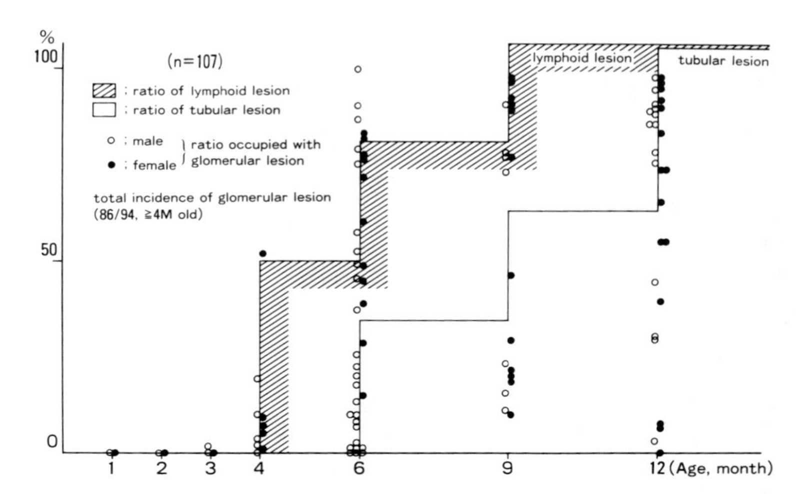

Fig. 2. The incidences of three major histopathological abnormalities are shown. The incidence of lymphoid follicle-like lesions (hatched area) and tubular lesions (open area) in various age group of NON mice are shown. The ratio of affected glomeruli with glomerular lesion in total glomeruli which was evaluated on LM by longitudinal section of the kidney in each mice is shown as a circle (open circle: male, closed circle: female, respectively). The incidence of three lesions were increased by aging.

Glomerular lesion

Massive deposits consisted of lipids and a proteinaceous substance within the capillary lumina was the major characteristic finding by LM (Fig. 3). This lesion was accompanied by mild mesangial cellular proliferation and mesangial matrix expansion. Therefore, this lesion resembled the diabetic nodular and/or exudative lesion (Fig. 4). However, the detailed nature of this lesion was completely different from that of human diabetic nephropathy. Since the deposits within capillary lumina had a growing up nature in situ, mesangiolysis was induced and ballooning of capillary lumina was seen as a result. The appearance of this lesion was gradually increased by aging and overall frequency was 80.4% (86/107). When the rate of this lesion was calculated in mice over the age of 4 months, it was 91.5% (86/94). This lesion was found with similar frequency in both sexes. Regarding the relationship between this lesion and IGT; it was same for the group with no IGT and that showing IGT (91% for both).

Various lipid staining methods showed positive results, thus indicating the existence of many different kinds of lipids (e.g., triglycerides, unsaturated cholesterol esters, phospholipids, unsaturated hydrophobic lipids, free cholesterol, and lipoproteins (Fig. 5).

Direct IF showed positive staining against IgG, A, and M at mesangial areas as well as along the glomerular capillary wall in both NON and ICR mice (Fig. 6). However, the degree was stronger in NON mice. IgM showed the strongest positive staining and positive staining was also found in the intracapillary deposits in NON mice.

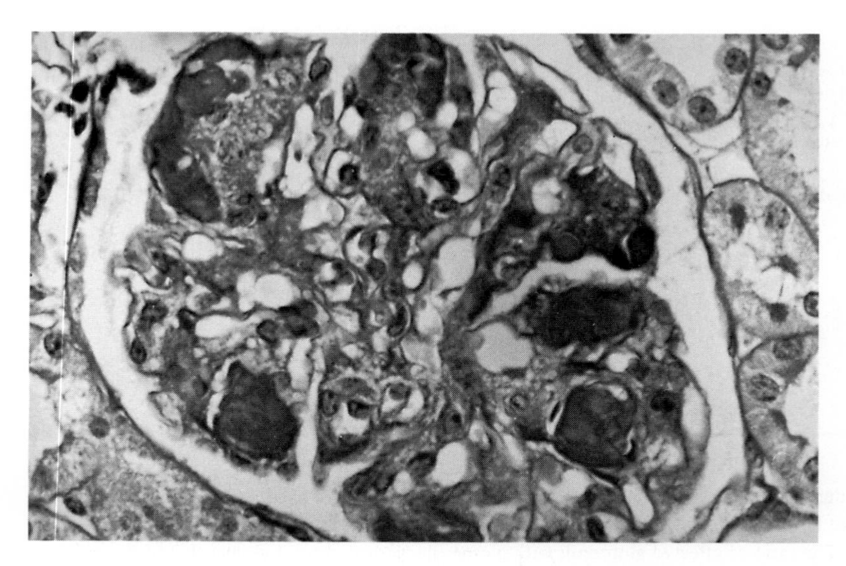

Fig. 3. Light microscopic finding of glomerulus in 6-month-old NON mice. Massive PAS positive deposits are observed within the glomerular capillary lumen. PAS $\times 200$.

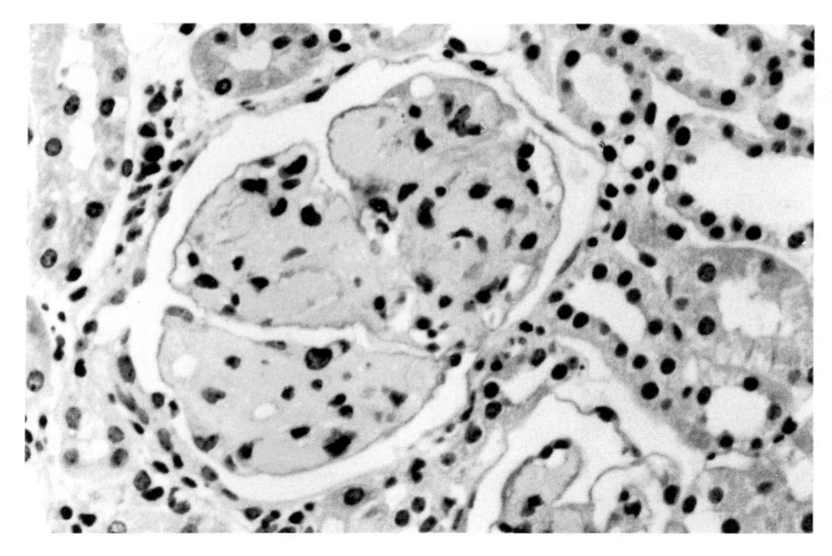

Fig. 4. Light microscopic findings of glomerulus of 6-month-old NON mice which shows a nodular lesion-like appearance. HE $\times$ 100.

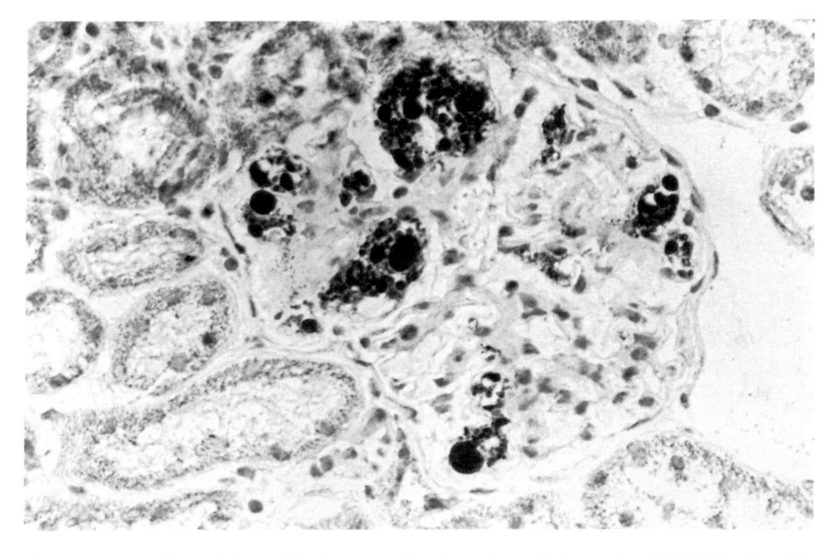

Fig. 5. Sudan III staining with frozen unfixed section of 6-month-old NON mice. The positive staining within glomerular capillar lumen certify the existence of lipids. $\times$ 100

Fig. 6. Immunofluorescence against anti-mouse IgM. Strongly positive finding was observed at mesangium as well as deposits within capillary lumen. $\times$ 100.

EM examination showed that the amorphous deposits occupied entire capillary lumina and these deposits induced not only mesangiolysis but also thinning of the glomerular basement membrane (Fig. 7). In the early stage, the deposits were seen at subendothelial space too (Fig. 8). In addition, the deposits were likely to be floating within glomerular capillary lumina at an early stage. These findings are likely to support the hypothesis that the massive deposits are formed and grow up in situ like a thrombus. Numerous spherical substances were also seen in these deposits by high magnification. These spherical substances were considered to represent lipids (Fig. 9).

Tubular lesion

The major finding in tubules was the dilatation of tubular lumina and this lesion was always found in high aged NON mice. Proximal as well as distal tubules were dilated and the dilatation of pelvic calyces and even hydronephrosis was observed macroscopically in severe cases. The tubular cells itself were almost intact.

Lymphoid lesion

The third characteristic finding were the lymphoid follicle-like structures around the renal arterioles. This lesion was observed from 4 months of age onwards, and corresponded well to the existence of glomerular lesion. Since lymphocytes infiltration around pancreatic islets was a specific finding in NOD mice, we carried out the

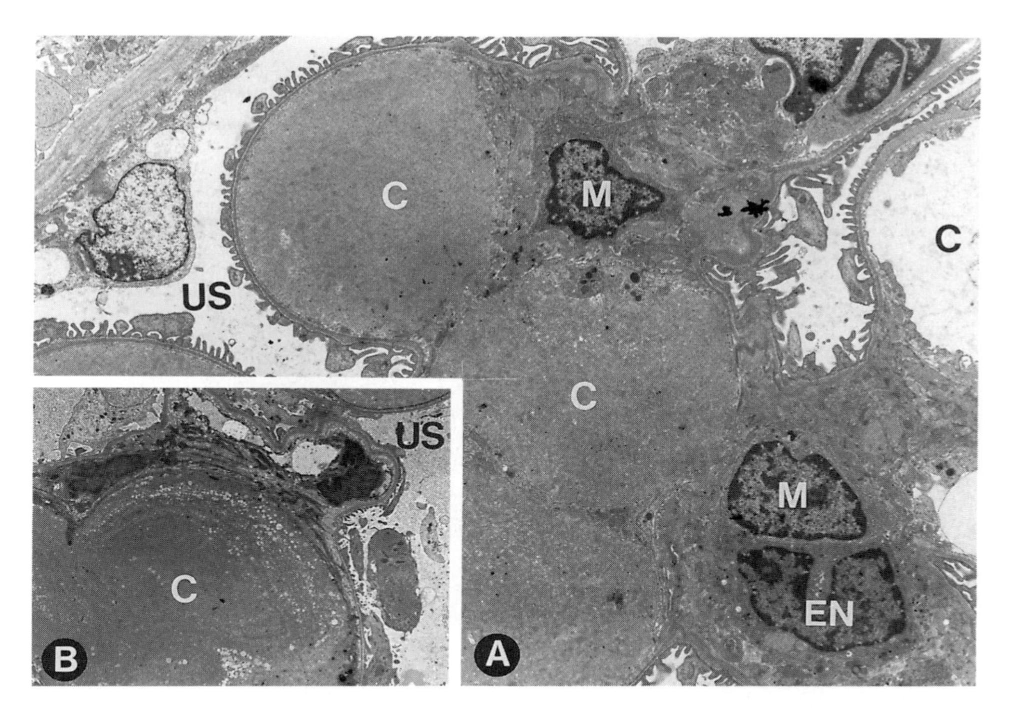

Fig. 7. Electron microscopic findings of 6-month old NON mice (A). Massive space-occupying proteinaceous deposits are seen entire of capillary lumen, These deposits cause mesangiolysis and thinning of glomerular basement membrane, ×4 000. [inset (B)]: The EM findings of the case of human lipoprotein glomerulopathy ×2 000) Abbreviations are: M: mesangial cell, EN: endothelial cell, C: capillary lumen, US: urinary space.

characterization of lymphocytes in the kidney tissue. Most of the accumulated lymphocytes were positive against Ly-1 and L3T4 antibodies (Fig. 10). These results suggest that the lymphocytes are mainly composed of helper T cells.

Histopathological study of various organs

Neither lipid deposits nor any thrombi were found in other organs. This result suggests that the deposits were formed exclusively in the kidney. Mild lymphocytes infiltration were found in the pancreas in some of the NON mice, however, the degree was mild and the structures of the islets were intact. Though the destruction of pancreatic islets by infiltrated lymphocytes was the major pathogenetic mechanism of NOD mice, the finding of lymphocytes infiltration in the pancreas of NON mice did not show any relationship with IGT.

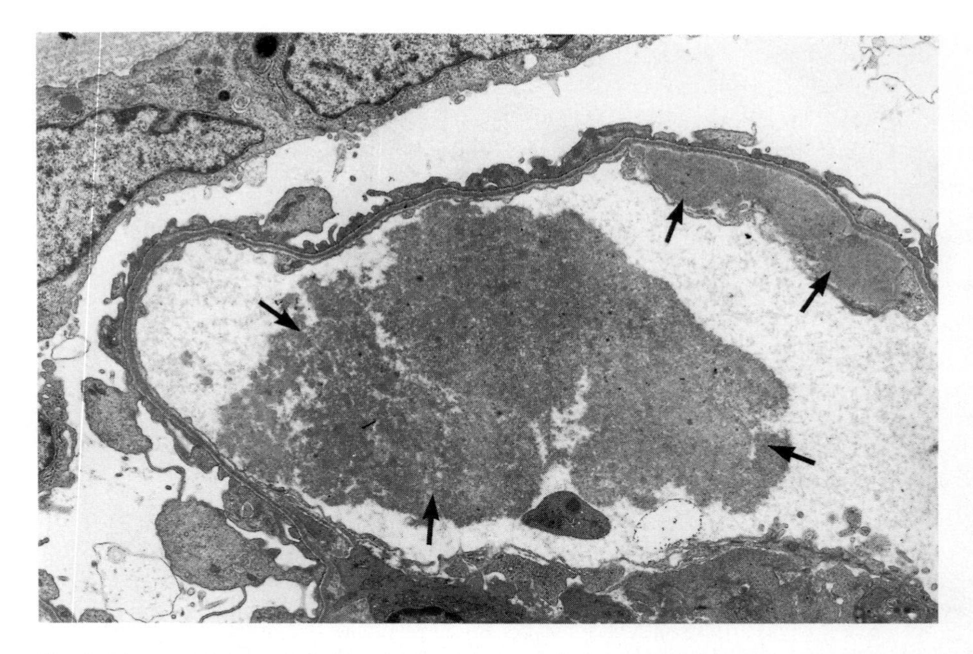

Fig. 8. Electron microscopic findings in the early stage of glomerular lesions obtained from 4-month-old NON mice. The deposits (arrow) are observed within glomerular capillary lumen as well as subendothelial space. ×2 600.

Serum immunoglobulin assay

Serum immunoglobulin levels were measured in both NON and ICR mice, and the values are shown in Table II. The values are likely to increase with age in both strains of mice, however, the values are always higher in the NON strain than in the ICR strain. Statistically significant differences were observed in the titer of IgG in all age groups.

Characterization of lymphocytes

The analysis of T-lymphocytes in spleen revealed no significant difference between NON and ICR mice. The mean values of the percentage of Lyt-2 positive cells (suppresor T cells) and L3T4 cells (helper T cells) were as follows: NON: 3.8%, 13.5%; ICR, 4%, 13.2%, respectively. These results mean that the systemic population of T-lymphocytes are same between two strains. Therefore, it is likely that the lymphoid lesion with predominant helper T cells is derived from some specific immunological disorder of NON mice against kidney tissues.

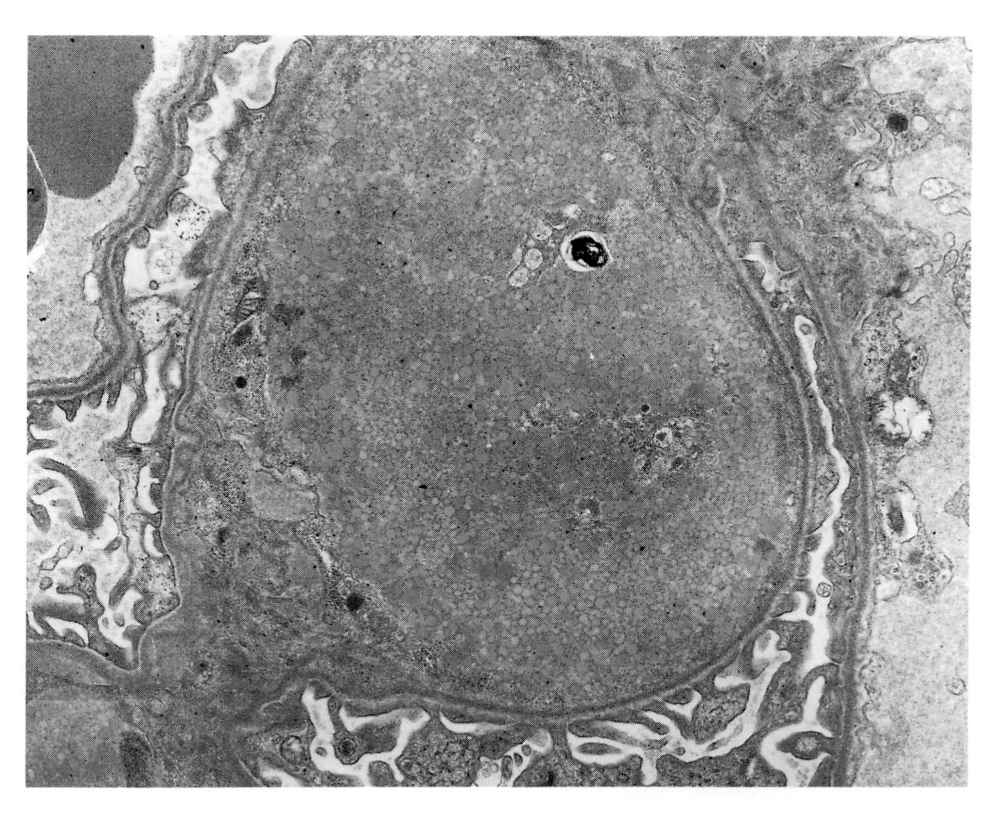

Fig. 9. Electron microscopic findings of deposits with high magnification. Numerous spherical substances which are considered to be lipids are shown. ×4 600.

TABLE II

The mean values of serum immunoglobulin concentration in various age classes of NON and ICR strains of mice. (Values are expressed as mean ± SE, unit: mg/dl)

Age (months)	IgG (mg/dl)		IgM (mg/dl)		IgA (mg/dl)	
	ICR	NON	ICR	NON	ICR	NON
3	219±35 (10)	528±66[b] (6)	12.7±1.2 (10)	18.1±1.4[a] (6)	121±17 (10)	175±21 (6)
4	–	563±94 (10)	–	16.0±1.2 (10)	–	539±75 (10)
6	370±45 (7)	680±44[b] (30)	20.7±2.4 (7)	23.3±1.2 (30)	245±24 (7)	426±60 (30)
9	–	962±111 (15)	–	34.6±2.9 (15)	–	718±223 (15)
12	–	1493±178 (15)	–	40.6±3.7 (15)	–	1233±380 (15)

[a] $P<0.05$
[b] $P<0.01$

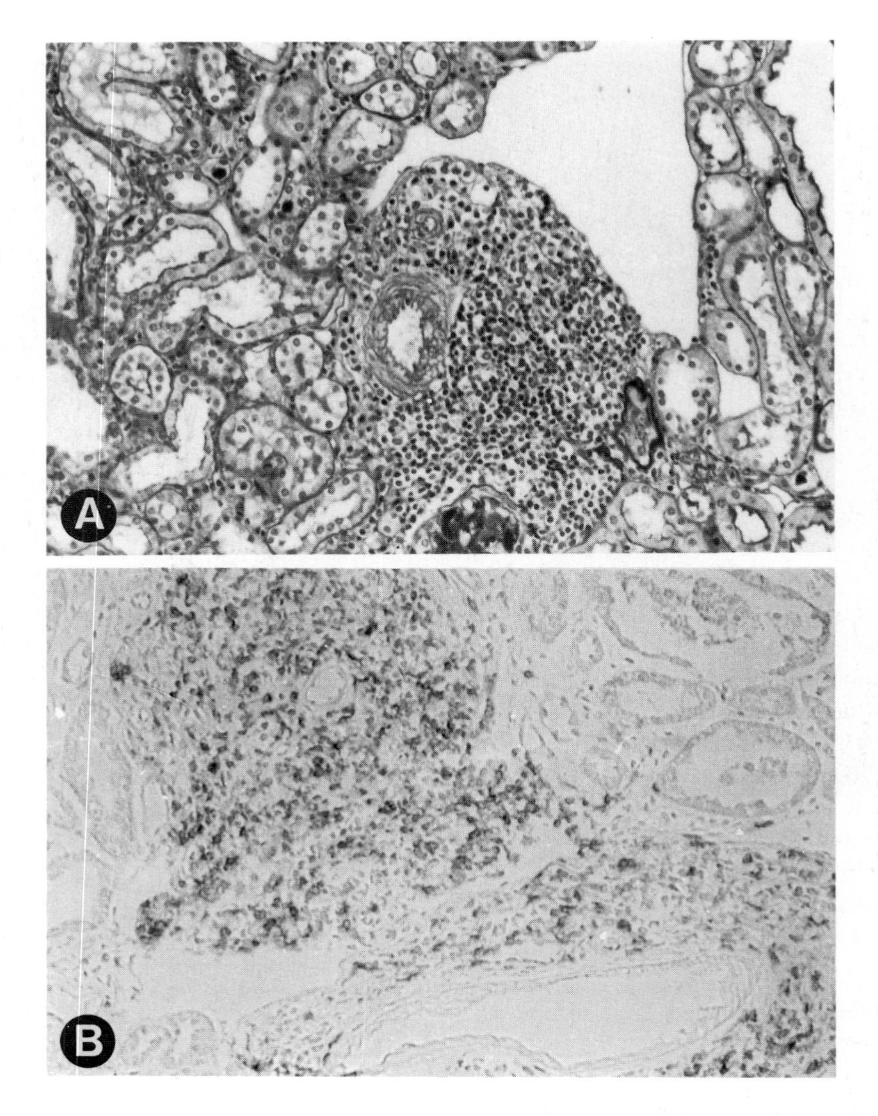

Fig. 10. The light microscopic finding of lymphoid lesion. Massive accumulation of lymphocytes are observed around arterioles (A). Positive staining of infiltrated lymphocytes against L3T4 is shown (B). (PAP-method) ×100.

Discussion

The NON strain of mice has been considered to be a useful diabetic animal model since it had IGT without destruction of the pancreatic islet [4–5]. We therefore started a histopathological survey of the kidney in this animal model. A unique glomerular lesion which resembled human diabetic nephropathy was found in this strain of mice with high incidence (91.5% over 4 months of age). However, the morphological characteristics were different in several points: (1) absence of diffuse thickening of glomerular basement membrane; (2) the deposits had the nature of thrombi contrary to that of plasmastasis in exudative lesions of diabetes; (3) the intensity of staining of this lesion was paler on HE staining; (4) the arrangement of mesangial cells was different from that of nodular lesions in diabetes. The latter usually shows an onion skin-like pattern on PAM staining and cell nuclei are arranged circumferentially. Meanwhile, the glomerular lesion in NON mice showed a lacework pattern on PAM staining and cell nuclei were scattered within the expanded mesangial area. In addition, there was no relationship between the frequency of glomerular lesion and IGT in NON mice. It was observed with the same frequency between the mice with IGT as without IGT. Thus, we consider that the glomerular lesion is specific to NON mice, and NON mice might have acquired some hereditary abnormalities in developing the glomerular lesion. However, this lesion is irrespective to the IGT which was another characteristic of NON mice.

Regarding the pathogenesis of this glomerular lesion, we found some interesting features in NON mice. NON mice showed higher values of serum immunoglobulins than the ICR strain of mice. Lymphoid lesion was also observed in the kidney of NON mice, and the accumulated lymphocytes were mainly composed of helper T cells. Since the appearance of a lymphoid lesion corresponded well to the formation of a glomerular lesion, we suppose that some immunological disarrangement might participate in the formation of glomerular lesion. This speculation is likely to be supported by the positive stainings against immunoglobulins by IF. Additional support would be obtained from the findings of NOD mice. NOD mice are the sibling strain of NON mice, and the pathogenesis of diabetes mellitus of NOD mice is, at present, explained by an autoimmune mechanism against pancreatic islets [8, 9]. Histologically, the pancreatic islets were destroyed by infiltrated lymphocytes in NOD mice, and most of the infiltrated lymphocytes were revealed to be helper T cells [10]. Genetic analysis revealed that three loci were required for the occurrence of lymphocytic infiltration in NOD mice, and NON mice posessed one of the genes present in NOD mice [11, 12]. Thus, we suppose that some genetically decided autoimmune mechanism might play a role in the appearance of glomerular lesions.

Concerning the glomerular lipid deposits, the important role of lipid deposition on the deterioration of renal functions has attracted attention [13, 14]. There have been some reports that demonstrated lipid depositon in glomeruli [15, 16]. However, these lipid depositions were restricted to mesangial areas and they were induced by either experimental nephrotic models or hyperlipidemic states by artificial feeding. The NON strain of mice is a unique animal model because it developed lipid deposition spontaneously without special treatment. There are many diseases which display lipid deposition in humans [17]. However, the histological characteristics of most of those diseases are completely different from that of NON mice. Human lipoprotein glomerulopathy is the only disease which shows similar glomerular morphological abnormalities. This disease was reported recently and was classified as a new category of renal disorder [18, 19]. The morphological characteristics of this disease are represented as lipid deposition in glomerular capillary lumina with capillary ballooning. In addition, the hereditary nature of this disease is now established from the finding of familial accumulation [20]. Thus, NON mice might share some of the characteristics with those of human lipoprotein glomerulopathy (Fig. 11).

Regarding tubular lesion, we observed it in all cases at the age of 12 months. Some of them showed the appearance of hydronephrosis. Therefore, NON mice should have the abnormality in tubules too. But this lesion is also considered to be unrelated with IGT.

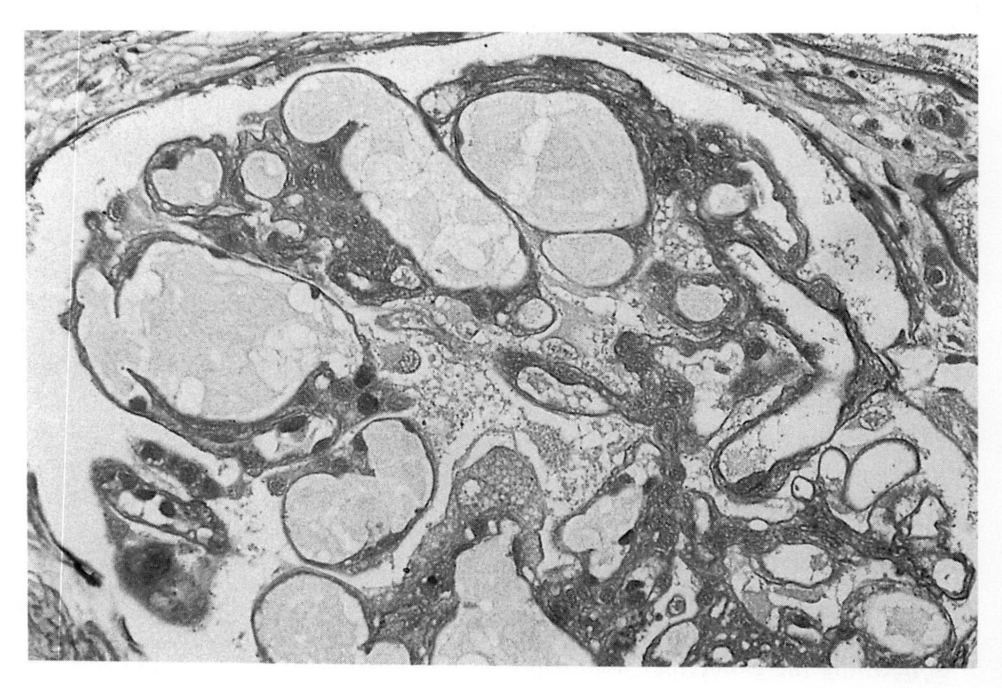

Fig. 11. Light microscopic finding of glomerulus in the case of human lipoprotein glomerulopathy. Similar characteristics with glomerular lesion of NON mice is seen. Masson-Trichrome, $\times 200$.

In conclusion, we found various renal lesions in NON mice. Among of them, glomerular lesion is unique. Though, we cannot find out any specific abnormalities of lipid metabolism in NON mice as far as we examined, this animal model would be important for investigating the role of lipid deposition in renal function. To elucidate the pathogenesis of human lipoprotein glomerulopathy, more intense study including lipid metabolism is necessary in the near future.

Summary

We found a unique glomerular lesion accompanied by a tubular lesion and a lymphoid follicle-like lesion in non-obese non-diabetic mice (NON). This strain of mice was developed together with a non-obese diabetic strain of mice (NOD) from an outbred ICR strain of mice. Since NON mice were reported to show impaired glucose tolerance, we carried out a renal histopathological study on this strain of mice with the aim of finding out whether these mice had diabetic lesion or not. The glomerular lesion which resembled that of human diabetic nephropathy was observed frequently, however, the detailed characteristics were different from that of human diabetic nephropathy. The results of the glucose tolerance test suggest that there is no relation between the occurrence of these lesions and IGT. The major findings were massive lipid deposition within glomerular capillary lumina and those space-occupying deposits were likely to cause mesangiolysis as well as thinning of glomerular basement membrane. Thus, glomerular ballooning, which was one of the characteristics of human lipoprotein glomerulopathy, was seen in NON mice. This strain of mice showed higher immunoglobulin levels in serum. In addition, the lymphocytes which infiltrated around the renal arterioles were revealed to be mainly composed of helper T cells by monoclonal antibody staining against L3T4. These findings suggest that these renal lesions were induced from some immunological disarrangements. Since glomerular lipid deposition has attracted attention as the major factor of deterioration of renal functions, this model may be a useful animal model in studying the role of lipids in renal diseases.

References

1 Valasquez MT, Kimmel PL, Michaelis OE. Animal models of spontaneous diabetic kidney disease. FASEB J 1990;4:2850–2859.

2 Leiter EH. The genetics of diabetes susceptibility in mice. FASEB J 1989;3:2231–2241.

3 Makino S, Kunimoto K, Muraoka Y, Mizushima Y, Katagiri K, Tochino Y. Breeding of a non-obese, diabetic strain of mice. Exp Anim 1980;29:1–13.

4 Tochino Y, Kanaya T, Makino S. Microangiopathy in the spontaneously diabetic nonobese mouse (NOD mouse) with insulitis. In: Abe H, Hoshi M, eds. Diabetic microangiopathy. Tokyo: University of Tokyo Press, 1983:423–432.

5 Kobayashi M, Ohgaku S, Maegawa H, Watanabe N, Takata Y, Shigeta Y. Insulin receptors in NOD mice. In: Tarui S, Tochino Y, Nonaka K, eds. Insulitis and type I diabetes. Tokyo: Academic Press, 1986;225–231.
6 Pears AGE. Lipids and lipoproteins. In: Pears AGE, ed. Histochemical theoretical and applied. Edinburgh, Churchill Livingstone, 1985;2:786–849.
7 Nakane PK, Pierce GB. Enzyme-labeled antibody for the light and electron microscopic localization of antigens. J Histochem Cytochem 1966;14:929–931.
8 Ogawa M, Maruyama T, Hasegawa T, Kanaya F, Tochino Y, Uda H. The inhibitory effect of neonatal thymectomy on the incidence of insulitis in non-obese diabetes (NOD) mice. Biomed Res 1985;6:103–105.
9 Wicker LS, Miller BJ, Coker LZ, McNally SE, Scott S, Mullen Y, Appel M. Genetic control of diabetes and insulitis in the nonobese diabetic (NOD) mouse. J Exp Med 1987;165:1639–1654.
10 Koike T, Itoh Y, Ishii T, Takabayashi K, Maruyama N, Tomioka H, Yoshida S. Preventive effect of monoclonal anti-L3T4 antibody on development of diabetes in NOD mice. Diabetes 1987;36:539–541.
11 Hattori M, Buse JB, Jackson A, Glimcher L, Dorf ME, Minami M, Makino S, Moriwaki H, Kuzuya H, Imura H, Strauss WM, Seideman JG, Eisenbarth GS. The NOD mouse: Recessive gene in the major histocompatibility complex. Science 1986;231:733–735.
12 Prochazka M, Leiter EH, Serreze DV, Coleman DL. Three recessive loci required for insulin-dependent diabetes in nonobese diabetic mice. Science 1987;237:286–289.
13 Moorhead JF, Chan MK, Nahas M, Varghese Z. Lipid nephrotoxicity in chronic progressive glomerular and tubulo-interstitial disease. Lancet 1982;ii:1309–1311.
14 Kashiske BL, O'Donnell MP, Cleary MP, Keane WF. Treatment of hyperlipidemia reduces glomerular injury in obese Zucker rats. Kidney Int 1988;33:667–672.
15 Diamond JR, Karnovsky MJ. Exacerbation of chronic aminonucleoside nephrosis by dietary cholesterol supplementation. Kidney Int 1987;32:671–678.
16 Al-Shebeb T, Frohlich J, Magil AB. Glomerular disease in hypercholesterolemic guinea pigs: A pathologic study. Kidney Int 1988;33:498–507.
17 Faraggiana T, Churg J. Renal lipidoses: A review. Human Pathol 1987;18:661–679.
18 Watanabe Y, Ozaki I, Yoshida F, Fukatsu A, Itoh Y, Matsuo S, Sakamoto N. A case of nephrotic syndrome with glomerular lipoprotein deposition with capillary ballooning and mesangiolysis. Nephron 1989;51:265–270.
19 Saito T, Sato H, Kudo K, Oikawa S, Shibata T, Hara Y, Yoshinaga K, Sakaguchi H. Lipoprotein glomerulopathy: glomerular lipoprotein thrombi in a patient with hyperlipoproteinemia. Am J Kidney Dis 1989;13:148–153.
20 Watanabe Y, Kamei N, Itoh Y, Yoshida F, Matsio S, Sakamoto N. Lipoprotein glomerulopathy. A new kind of glomerular disease. Clinic All-Round (Tokyo). 1989;38:826–830.

CHAPTER 13

Renal pathological findings and abnormal GTT

HIDEKI WAINAI[a], TARO MARUYAMA[b], IZUMI TAKEI[a], KUNIZO KATAOKA[a], TAKAO SARUTA[a], KENTARO OGATA[c]

[a]*Division of Endocrinology and Internal Medicine, Keio University School of Medicine, Shinjuku-ku, Tokyo, Japan*

[b]*Division of Internal Medicine, Social Insurance Saitama Chuo Hospital, 4-9-3 Kitaurawa, Urawa-shi, Saitama, Japan*

[c]*Division of Pathology, Keio University School of Medicine, University School of Medicine, Shinjuku-ku, Tokyo, Japan*

Current concepts of a new animal model: The NON mouse
Edited by N. Sakamoto, N. Hotta and K. Uchida

Contents

Introduction

Non-obese, non-diabetic (NON) mice are a subline of non-obese diabetic (NOD) mice. Non-obesity, moderate glucose intolerance, low pancreatic insulin content and the absence of insulitis are characteristic of this subline. Therefore, it is considered to be a new animal model for Type 2 diabetes. In order to evaluate the adequacy of NON mice as a model for human diabetes, we performed intra-peritoneal glucose tolerance testing (IP-GTT) and observed renal histology for up to 72 weeks. We were particularly interested in the course of renal deterioration as it compared to the development of human diabetic nephropathy.

Materials and methods

NON mice were obtained from the Aburahi Laboratories, Shionogi Pharmaceutical Co., Shiga, Japan. IP-GTT (2 g/kg body weight) was performed after 4 hours of fasting in male mice from 8 to 80 weeks of age and in female mice from 8 to 67 weeks of age. Animals were sacrificed at 8, 24, 38 or 72 weeks, and their kidneys were removed for histological study. The tissues were studied by light microscopy (H-E, PAS or PAM staining) and electron microscopy.

Results

Glucose intolerance

At 8 weeks of age, NON mice manifest significant glucose intolerance compared with control ICR mice. Male mice have more marked abnormalities than female mice. There was also considerable individual variation in glucose intolerance. Glucose intolerance gradually improved with age in both males and females (Fig. 1 and 2).

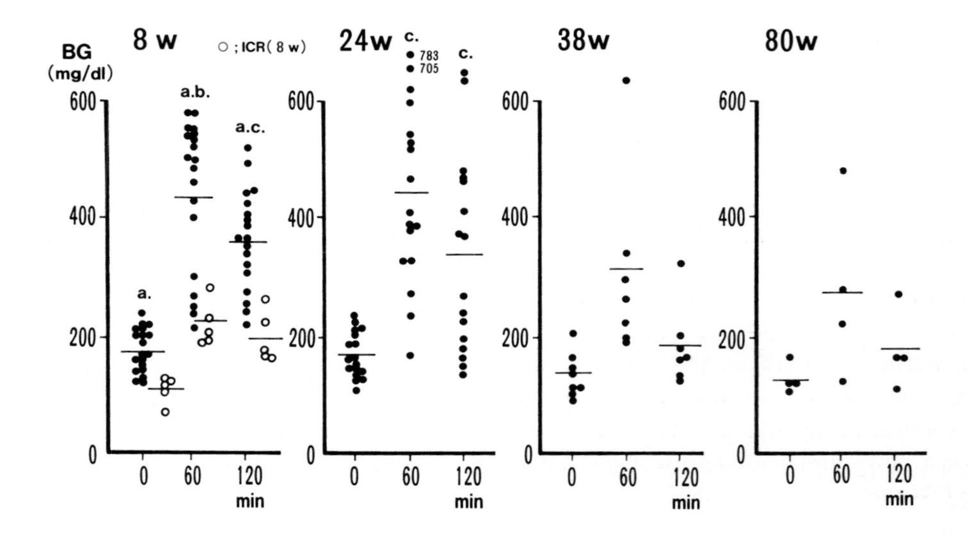

a. ; p<0.01 vs. ICR b. ; p<0.05 vs. female. c. ; p<0.01 vs. female.

Fig. 1. Time course of IP-GTT in male NON mice.

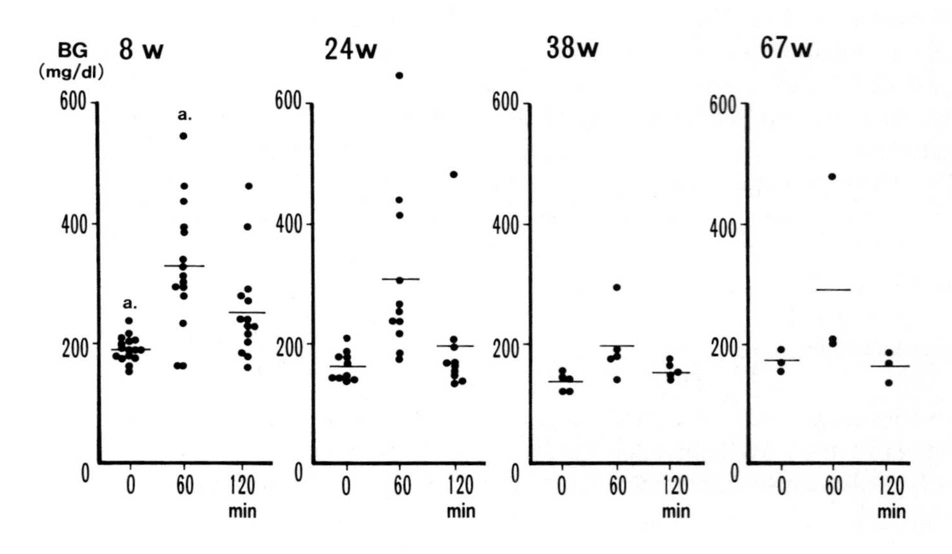

a. ; p<0.01 vs. ICR.

Fig. 2. Time course of IP-GTT in female NON mice.

Renal histology

At 8 weeks of age, some of the glomeruli showed mesangiolysis and microaneurysmal dilatation (Fig. 3). At 38 weeks of age, the renal glomeruli exibited marked morphological changes. The capillary lumina were dilated and filled with PAS-positive thrombus-like material (Fig. 4). Electron microscopically, this thrombus-like material consisted of small lipid-like droplets. Dense deposition within the mesangium and glomerular basement membrane and thinning of the basement membrane were also seen (Fig. 5). Minimal accumulation of small lipid-like droplets was already seen in the early stage of the lesion. These changes progressed with age, and at 72 weeks sclerotic lesions with mesangial proliferation were evident. At this age, some of the glomeruli showed crescent formation and adhesion to Bowman's capsule (Fig. 6). Another histological change, dilatation of tubules in the deep cortex, was seen only in females (Fig. 7), whereas interstitial cell infiltration was seen in both males and females.

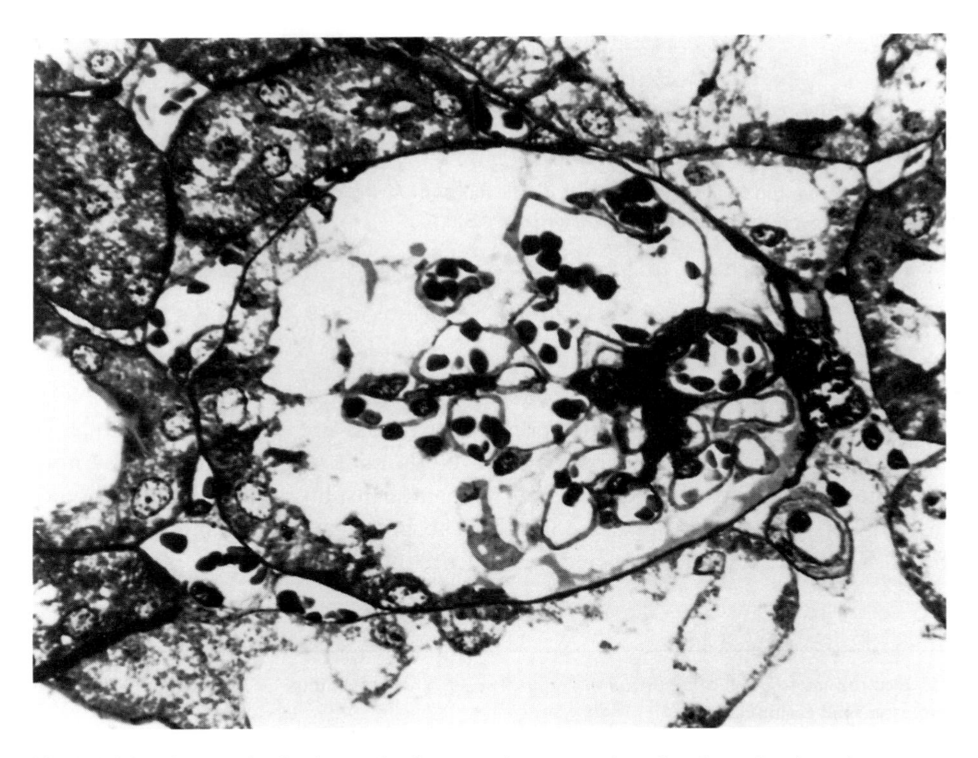

Fig. 3. Light micrograph of a glomerulus from a male mouse at 8 weeks of age, showing microaneurysmal dilatation of the glomerular vessels (PAM).

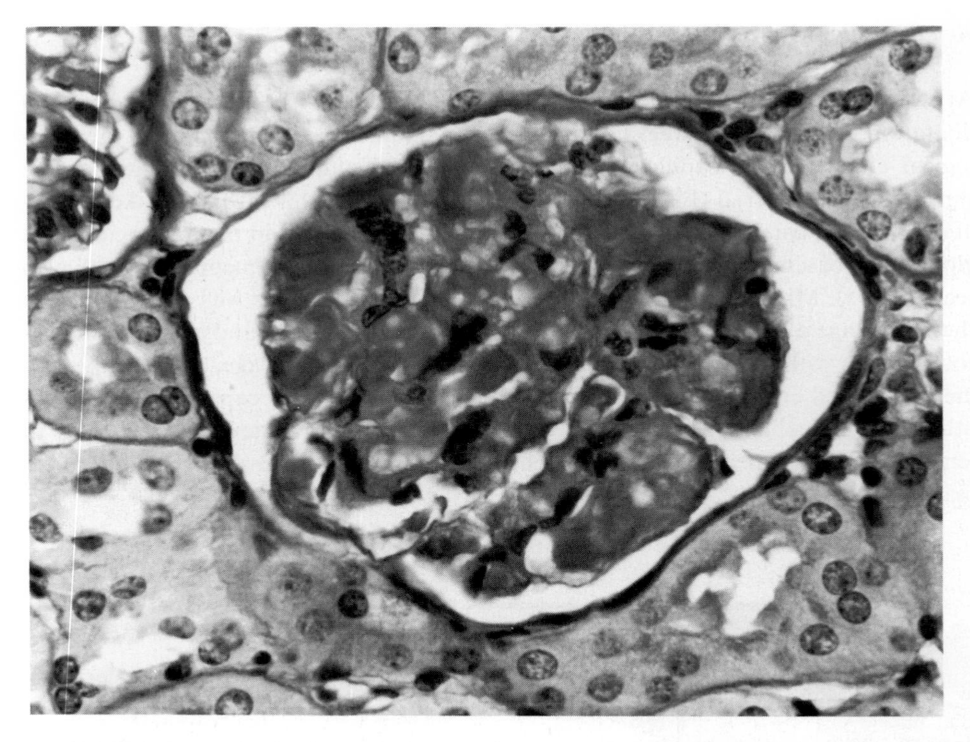

Fig. 4. Light micrograph of a glomerulus from a male mouse at 38 weeks. The capillary lumina are dilated and filled with PAS-positive thrombus-like material (PAS).

Discussion

NOD and NON mice were derived from the CTS strain of Jcl/ICR mice [1]. NOD mice develop diabetes with insulitis spontaneously and are considered a model of human Type 1 diabetes. On the other hand, NON mice are characterized by non-obesity, moderate glucose intolerance, low pancreatic insulin content and the absence of insulitis [2]. Therefore, they are considered to be a new animal model of type 2 diabetes. However, Kano [3] reported that glucose intolerance in NON mice gradual-

Fig. 5. Electron micrograph of a portion of a glomerulus from a male mouse at 38 weeks ($\times 2480$). Note the extensive lipid accumulation.

Fig. 6. Endstage glomerular lesion (72 weeks). Segmental sclerosis with crescent formation and adhesion to Bowman's capsule is seen (PAM).

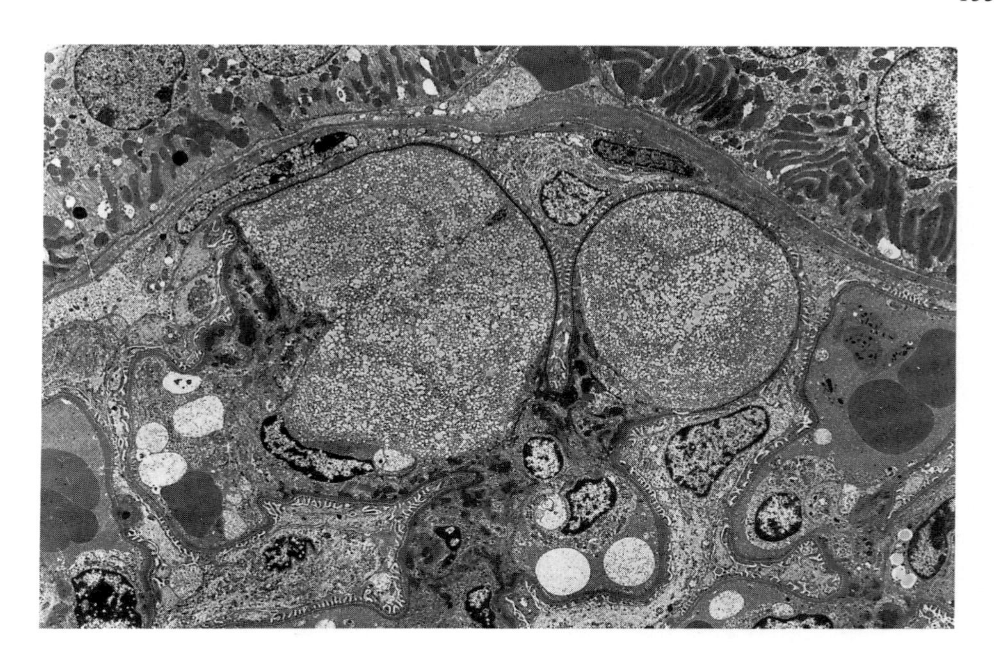

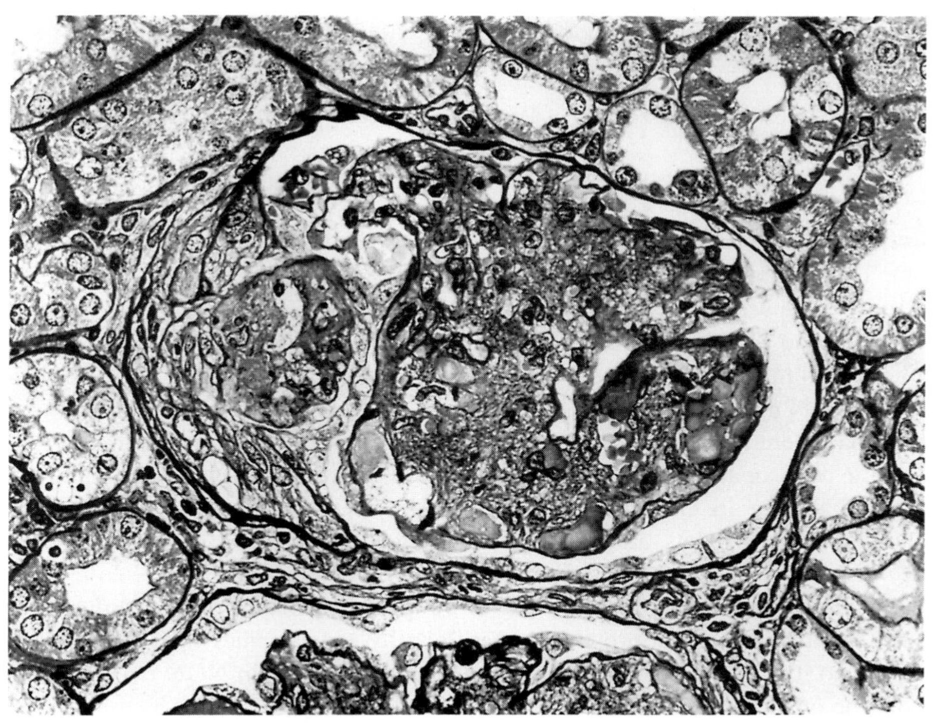

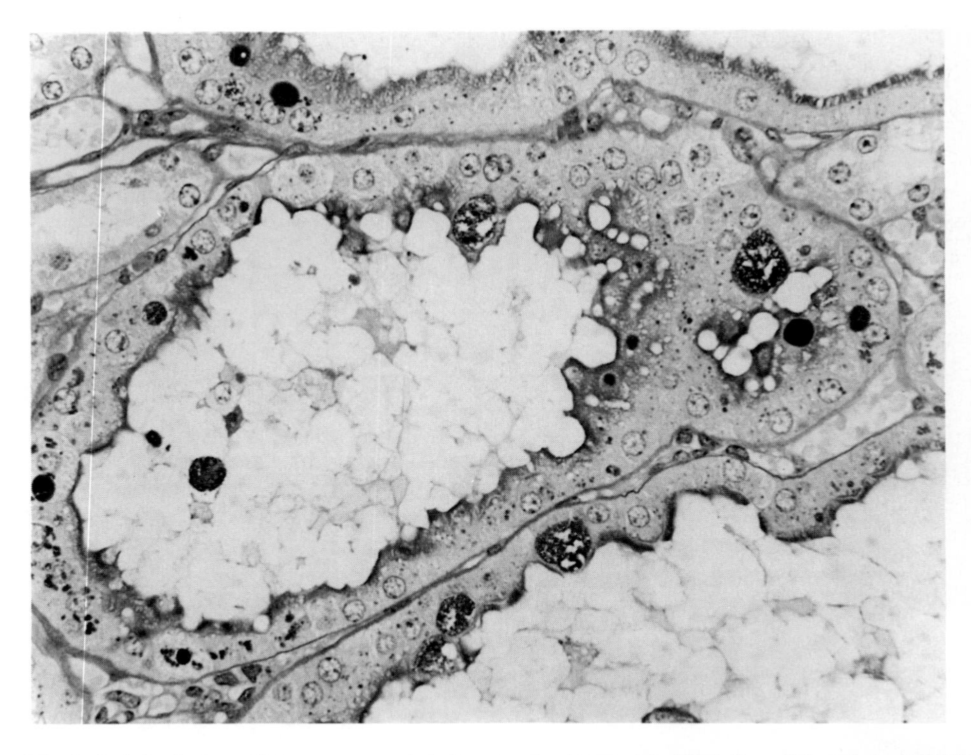

Fig. 7. Tubular lesion which was seen only in female mice (24 weeks). Dilatation of the tubule and PAS-positive droplets are seen (PAS).

ly disappeared with age, which is rare in humans. Our data are consistent with their report. The mechanism underlying this phenomenon in NON mice is unknown. In our study, there was also considerable individual variation in glucose intolerance in both males and females. Further investigation is necessary in order to determine whether NON mice may serve as a model for human Type 2 diabetes. The renal histology of NON mice was not typical of diabetic nephropathy. The thickening of glomerular basement membranes and nodular sclerotic lesions which are characteristic of diabetic nephropathy were not observed even at 72 weeks. The glomerular lesions seen in NON mice are characterized by mesangiolysis, resulting from accumulation of small lipid-like droplets, and resemble lipoprotein glomerulopathy, a rare familial kidney disease which has recently been reported [4, 5]. With regard to renal lesions, NON mice are considered to be a model of lipoprotein glomerulopathy rather than diabetic nephropathy. Lipoprotein glomerulopathy accompanies lipid metabolism disorders, so it will be necessary to examine lipid metabolism in NON mice. In conclusion, the time course of glucose intolerance and the change in renal histology seen in NON mice are unique. Furthermore, the renal histology is not typical of the renal complications seen in human diabetes.

References

1 Makino S, Kunimoto K, Muraoka Y, Mizushima Y, Katagiri K, Tochino Y. Breeding of a non-obese, diabetic strain of mice. Exp Anim 1980;29:1–13.
2 Ohgaku S, Morioka H, Yano S, Yamamoto H, Okamoto H, Tochino Y. NON mice: a new animal model for non-obese NIDDM. Diabetes Res Clin Pract 1985;Suppl. 1:S415.
3 Kano Y. Insulin secretion in NOD (non-obese diabetic) and NON (non-obese non-diabetic) mouse. J Kyoto Pref Univ Med 1988;97:295–308.
4 Watanabe Y, Ozaki I, Yoshida F, Fukatsu A, Itoh Y, Matsuo S, Sakamoto N. A case of nephrotic syndrome with glomerular lipoprotein deposition with capillary ballooning and mesangiolysis. Nephron 1989;51:265–270.
5 Saito T, Sato H, Kudo K, Oikawa S, Shibata T, Hara Y, Yoshinaga K, Sakaguchi H. Lipoprotein glomerulopathy: glomerular lipoprotein thrombi in a patient with hyperlipoproteinemia. Am J Kid Dis 1989;13:148–153.

CHAPTER 14

Renal lesions and urinary bladder carcinogenesis

SATORU MORI[a,b], TAKASHI MURAI[a,c], TADAO OHHARA[a], MOTOKO HOSONO[a], YASUYOSHI TAKEUCHI[a], SUSUMU MAKINO[a], YOSHIYUKI HAYASHI[a] and SHOJI FUKUSHIMA[c]

[a]*Aburahi Laboratories, Shionogi Research Laboratories, Shionogi and Co., Ltd., Koka, Shiga, Japan*
[b]*First Department of Pathology, Nagoya City University Medical School, Kawasumi, Mizuho, Nagoya, Japan*
[c]*First Department of Pathology, Osaka City University Medical School, Asahi, Abeno, Osaka, Japan*

Current concepts of a new animal model: The NON mouse
Edited by N. Sakamoto, N. Hotta and K. Uchida

Contents

Introduction

In the lower urinary tract, experimental animal models for urinary bladder carcinogenesis have been established, and the biological characteristics of the carcinomas have been extensively investigated [1–4]. The results have led to developments in diagnosis and chemotherapy of early stage urinary bladder carcinoma [5].

Many clinical reports indicate that diagnosing renal pelvic carcinomas at an early stage remains a difficult problem. By the time of diagnosis, most renal pelvic carcinomas have already invaded the kidney and/or the surrounding tissues [6–8]. Moreover, several epidemiological and clinical reports have emphasized the possibility that large amounts of analgesics containing phenacetin, antipyrene and caffeine for long periods result in induction of renal pelvic carcinoma, occasionally with urinary bladder carcinoma [9–14]. In carcinogenicity study, Sprague–Dawley (SD) rats given large amounts of phenacetin for long periods had low to moderate incidences of the urinary bladder and/or renal pelvic carcinomas [15–21]. B6C3F1 mice given the same treatment had low incidences of carcinomas of the renal pelvis and urinary bladder [22]. Consistent experimental animal models of renal pelvic carcinogenesis, however, have not been established.

A previous study suggested that ligation of the ureter before administration of N-butyl-N-(4-hydroxybutyl)nitrosamine (BBN), a well-known urinary bladder carcinogen, changes target organs from the urinary bladder alone to the renal pelvis and urinary bladder [23]. A similar phenomenon was observed with needle-puncture of the kidney [18, 19]. These procedures may produce stagnant urine flow in the renal pelvis, increasing exposure to the carcinogen. Recently, administration of the combination of uracil and BBN produced a few renal pelvis carcinomas [24, 25]. Since uracil induced calculi in the urinary bladder, ureter and renal pelvic [26–30], secondary hydronephrosis results. The hydronephrosis may also result in stagnant urine in the renal pelvis and ureter containing metabolites of BBN. The possibility therefore existed that a mutant strain of rodent having spontaneous hydronephrosis exposed to a urothelial carcinogen might be a suitable animal model for renal pelvic carcinogenesis.

We have found that NON/Shi mice have approximately 10 to 30% incidence of unilateral spontaneous hydronephrosis (Fig. 1) [31]. NON/Shi mice were derived

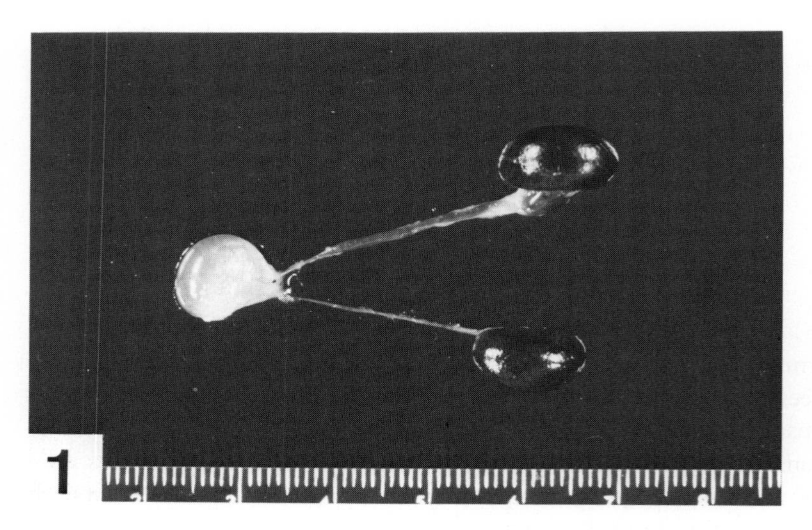

Fig. 1. Hydronephrosis and hydroureter of the right kidney and ureter in a male NON/Shi mouse at 45 wk of age.

from the ICR mouse strain and are a sister strain of NOD/Shi mice [32] which are an excellent animal model of Type I diabetes mellitus [33–36]. Moreover, NON/Shi mice have been discussed as a suitable strain for an experimental model of Type II diabetes mellitus [37–39]. However, the situation of NON/Shi mice with respect to renal pelvic and urinary bladder carcinogenesis by BBN has not been ascertained.

We evaluated whether NON/Shi mice have susceptibility for BBN-induced renal pelvic and urinary bladder carcinogenesis. The experiments suggest that NON/Shi strain is a suitable model animal for renal pelvic carcinogenesis.

Induction of renal pelvic carcinoma by BBN

The purpose of the present study was to examine the susceptibility of NON/Shi mice to renal pelvic and urinary bladder carcinogenesis by BBN.

Twenty-three NON/Shi (group 1), 17 DS/Shi (group 2) (Aburahi Laboratories, Shionogi & Co., Ltd.) and 19 B6C3F1 (group 3, Charles River Japan, Inc., Shiga, Japan) strains of male mice, 6 weeks old, were used. DS/Shi and B6C3F1 mice had no spontaneous hydronephrosis. Animals were given CA-1 diet (Japan Clea Ltd., Osaka, Japan) and drinking water containing 0.05% BBN (Tokyo Kasei Kogyo Co., Tokyo, Japan) for 12 wk, followed by water without chemicals until experiment wk 20. The mice were killed and the kidneys, urinary bladder, lung and abdominal organs with metastases were removed and fixed in formalin (pH 7.4) for histopathological examination. The epithelial lesions of the urinary bladder were classified into three categories: hyperplasia, papilloma and carcinoma as described previously [4, 40]. Also, these categories were applied to epithelial lesions of the renal pelvis [4]. Distribution of the carcinomas in the renal pelvis and urinary bladder were classified into categories: (A) unilateral renal pelvic carcinoma alone; (B) unilateral renal pelvic carcinoma and urinary bladder carcinoma; (C) bilateral renal pelvic carcinomas and urinary bladder carcinoma; and (D) urinary bladder carcinoma alone, modified from previously described categories [41].

Since 7 NON/Shi mice of group 1 died from renal pelvic and/or urinary bladder carcinomas during experimental wk 11 to 20, group 1 showed the lowest survival ratio (52%) compared to groups 2 (65%) and 3 (84%). Body weight gains were the same among the three groups. Daily BBN intake showed no differences between groups 1, 2 and 3.

Macroscopically, 7 NON/Shi mice (group 1) had large nodular tumors in the right renal pelvis (Fig. 2), whereas the other two strains of mice did not. Urinary calculi were not found in any mice of groups 1–3.

The histopathological lesions of the renal pelvis and urinary bladder are summarized in Table I. NON/Shi mice given BBN (group 1) had a 32% incidence of renal pelvic carcinomas (Fig. 3), whereas the other two strains (groups 2 and 3) had no renal pelvic carcinomas. One of the renal pelvic carcinomas metastasized to the lung (Fig. 4).

Incidences of urinary bladder carcinomas in groups 1–3 were 86, 94 and 95%, re-

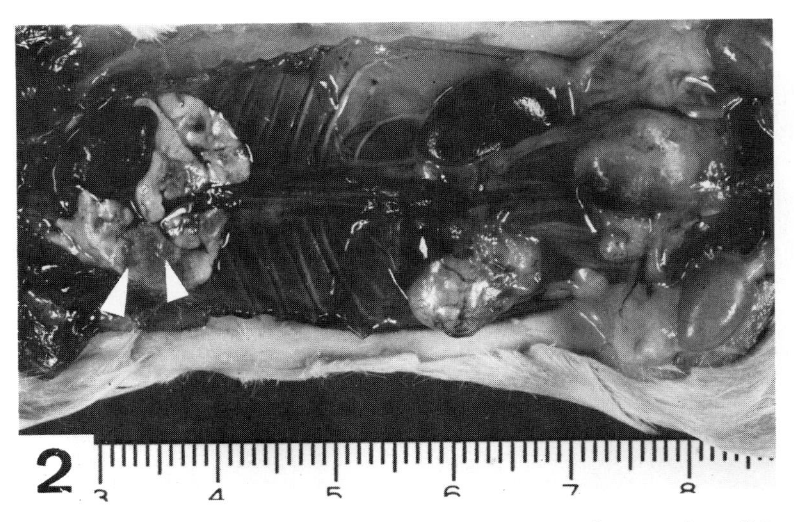

Fig. 2. Right renal tumor, urinary bladder tumors, and several metastatic small tumor nodules (arrow) are observed in a male NON/Shi mouse of group 1 (Section 2).

TABLE I

Mouse strain differences in BBN-induced renal pelvic and urinary bladder carcinogenesis

Group	Strain	Effective number of mice	Incidences of carcinomas		Distribution of categories[a]			
			Renal pelvis No.[b] (%)	Urinary bladder No. (%)	A No. (%)	B No. (%)	C No. (%)	D No. (%)
1	NON/Shi	22	7(32)	19(86)	2(9)	5(23)	0	14(64)
2	DS/Shi	17	0	16(94)	0	0	0	16(94)
3	B6C3F1	19	0	18(95)	0	0	0	18(95)

[a]Category: A, Unilateral renal pelvic carcinoma alone; B, Unilateral renal pelvic and urinary bladder carcinomas; C, Bilateral renal pelvic and urinary bladder carcinomas; D, Urinary bladder carcinoma alone.
[b]Number of mice with lesion.

spectively. Urinary bladder carcinomas in groups 1, 2 and 3 metastasized to abdominal organs in 1, 1 and 2 mice of the these groups, respectively.

Distribution of the categories for carcinomas of the renal pelvis and urinary bladder were 2 and 5 of 22 NON/Shi mice (group 1) in categories A and B, respectively. Category C was not presented in any of the groups. Incidences of category D in groups 1, 2 and 3 were 64, 94 and 95%, respectively.

No papillomas were found in the renal pelvis or urinary bladder in any group. Since urinary bladder carcinomas produced secondary hydronephrosis, primary hydronephrosis was not observed in any mice of groups 1–3.

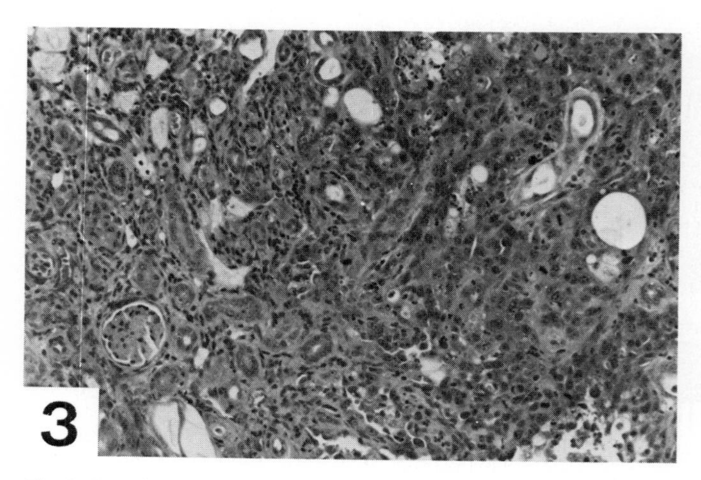

Fig. 3. Invasive transitional cell carcinoma in right kidney of a male NON/Shi mouse of group 1 (Section 2). H.E. stain, × 200.

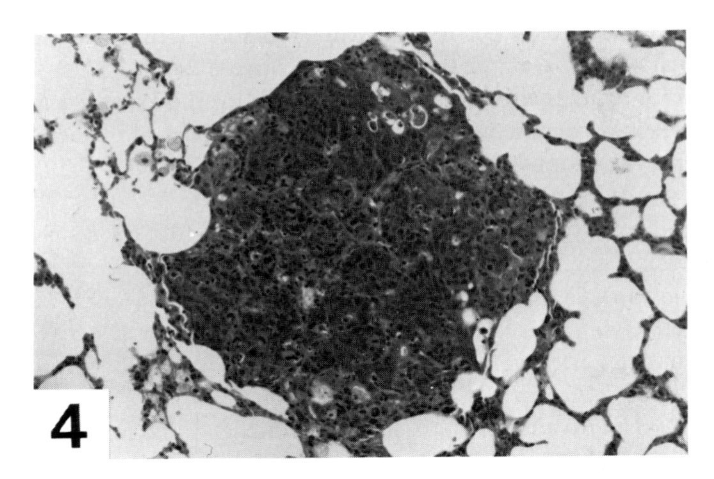

Fig. 4. Alveolar sac and pulmonary artery have metastatic transitional cell carcinoma from a renal pelvic carcinoma (Fig. 3) of a category B-bearing NON/Shi mouse in group 1 (Section 2). H.E. stain, × 200.

The results indicate that NON/Shi mice are susceptible to induction of the renal pelvic carcinomas by BBN, but DS/Shi and B6C3F1 mice, lacking spontaneous hydronephrosis, are not.

Influence of age and sex on BBN-induced renal pelvic carcinogenesis in NON/Shi mice

Since male NON/Shi mice given 0.05% BBN for 12 wk are susceptible to renal pelvic carcinogenesis, we investigated whether age and sex influenced the process.

Male and female NON/Shi mice were used: 80 males and 77 females, 6 wk old (groups 1 and 4, respectively), 14 males and 16 females, 15 wk old (groups 2 and 5), and 30 males and 48 females, 45 wk old (groups 3 and 6). In the first 12 wk of the experiment, drinking water containing 0.05% BBN was given to the mice, and then they were given water for 13 wk without added chemicals.

Survival was low in groups 3 and 6, intermediate in groups 1, 2 and 5, and high in group 4. A similar trend was observed in time to 50% mortality and body weight gain. Changes in survival ratio and body weight gain were correlated with age but not with sex. Daily BBN intake in groups 1, 3 and 6 was similar, but that in groups 2, 4 and 5 was lower.

Macroscopically, unilateral kidney tumors of groups 1–6 were found in 18, 5, 2, 6, 1 and 12 mice, respectively. The majority of tumors were present in the right kidney. Bilateral tumors in kidneys in groups 1, 3 and 6 were present in 1, 1 and 2 mice, respectively. Urolithiasis was not found in any mouse.

Incidences of carcinomas in the renal pelvis and urinary bladder are shown in Table II. Incidences of renal pelvic carcinomas in groups 1–6 were 24, 38, 37, 8, 6 and 29%, respectively. Several renal pelvic carcinomas in all groups metastasized to liver (Fig. 5), aorta, pancreas and/or mesentery lymph nodes: incidences of metastasizing renal pelvic carcinomas in groups 1–6 were 21% (4/19), 0% (0/5), 30% (3/10), 17% (1/6), 0% (0/1) and 7% (1/14), respectively.

Incidences of urinary bladder carcinomas in groups 1–6 were 91, 100, 67, 89, 94 and 90%, respectively. Metastasizing urinary bladder carcinomas to lung or abdominal organs occurred in all groups: 11% (8/73), 15% (2/13), 17% (3/18), 9% (6/68), 7% (1/15) and 7% (3/43), respectively.

Distribution of tumors in categories were: category A, 26% (7/27) mice of group 3; category B, 38% (5/13) mice of group 2 > 23% (18/80) and 25% (12/48) mice of groups 1 and 6 > 7% (2/27), 8% (6/76) and 6% (1/16) mice of groups 3, 4 and 5, respectively; category C, 1% (1/80), 4% (1/27) and 4% (2/48) mice of groups 1, 3 and 6, respectively; and category D, 68% (54/80), 62% (8/13), 56% (15/27), 82% (62/76), 88% (14/16) and 60% (29/48) mice of groups 1 to 6, respectively.

The results demonstrate that age and sex strongly influence susceptibility to BBN-induced malignant renal pelvic carcinogenesis in NON/Shi mice; old males are the most susceptible. Old male mice had a 30% incidence of metastases.

TABLE II

Effects of age and sex on BBN-induced renal pelvic and urinary bladder carcinogenesis in NON/Shi mice

Group	Age (in weeks)	Effective number of mice	Incidence of carcinomas		Distribution of categories[a]			
			Renal pelvis No.[b] (%)	Urinary bladder No. (%)	A No. (%)	B No. (%)	C No. (%)	D No. (%)
Male								
1	6	80	19(24)	73(91)	0	18(23)	1(1)	54(68)
2	15	13	5(38)	13(100)	0	5(38)	0	8(62)
3	45	27	10(37)	18(67)	7(26)	2(7)	1(4)	15(56)
Total		120	34(28)	104(87)	7(6)	25(21)	2(2)	77(64)
Female								
4	6	76	6(8)	68(89)	0	6(8)	0	62(82)
5	15	16	1(6)	15(94)	0	1(6)	0	14(88)
6	45	48	14(29)	43(90)	0	12(25)	2(4)	29(60)
Total		140	21(15)	126(90)	0	19(14)	2(1)	105(75)

[a]Category: A, Unilateral renal pelvic carcinoma alone; B, Unilateral renal pelvic and urinary bladder carcinomas; C, Bilateral renal pelvic and urinary bladder carcinomas; D, Urinary bladder carcinoma alone.
[b]Number of mice with lesion.

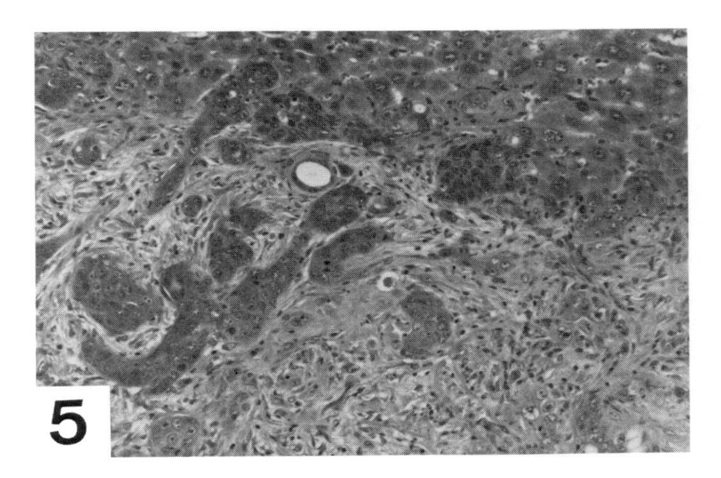

Fig. 5. Liver has metastatic transitional cell carcinoma from a renal pelvic carcinoma of a category B-bearing NON/Shi mouse in group 3 (Section 3). H.E. stain, × 200.

Co-carcinogenic effects of uracil on BBN-induced renal pelvic and urinary bladder carcinogenesis in male NON/Shi mice compared to male ICR mice

Since NON/Shi mice are sensitive to BBN-induced renal pelvic and urinary bladder carcinogenesis, we investigated whether they were susceptible to the co-carcinogenic effects of uracil on renal pelvic carcinogenesis.

Forty NON/Shi and 40 Crj : CD-1 (ICR) (Charles River Japan, Inc., Shiga, Japan) strains of male mice, 6 wk old, were given CA-1 diet, and divided into two groups of 20 mice each (NON/Shi mice, groups 1–2; ICR mice, groups 3–4). In the first 8 wk of the experiment, groups 1 and 3 were given CA-1 diet containing 3% uracil (99.8% pure, from Wako Pure Chemical Industries, Ltd., Osaka, Japan) and drinking water containing 0.05% BBN, and the matched control groups 2 and 4 were given drinking water containing 0.05% BBN and basal CA-1 diet. After the 8 wk of chemical administration, water and diet containing no chemicals were available to the mice until experimental wk 20.

Macroscopically, renal pelvic tumors occurred in 3 and 2 NON/Shi mice of groups 1 and 2, respectively, whereas none were observed in any ICR mice of groups 3 or 4. Urolithiasis was not found in any of the mice.

Histopathological lesions of the renal pelvis and urinary bladder are summarized in Table III. Incidences of renal pelvic carcinomas were low, less than 16%, in groups 1 and 2 of NON/Shi mice, whereas it was zero in groups 3 and 4 of ICR mice. Administration of uracil increased BBN-induced hyperplasia in the renal pelvis of both mouse strains. On the other hand, urinary bladder carcinomas were decreased in

TABLE III

Co-carcinogenic effects of uracil on BBN-induced renal pelvic and urinary bladder carcinogenesis in NON/shi and ICR strains of male mice

Group	Strain	Chemicals		Effective number of mice	Renal pelvis		Urinary bladder	
		BBN	Uracil		Hyperplasia No. [a] (%)	Carcinoma No. (%)	Hyperplasia No. (%)	Carcinoma No. (%)
1	NON/Shi	+	+	19	13(68)	3(16)	19(100)	2(11)
2	NON/Shi	+	−	19	4(21)	2(11)	18(95)	9(47)
3	ICR	+	+	15	9(60)	0	13(87)	0
4	ICR	+	−	18	1(6)	0	12(67)	2(11)

[a]Number of mice with lesion.

groups 1 and 3, when given BBN plus uracil, compared to the matched control groups 2 and 4 when given BBN alone. Moreover, co-administration of uracil did not increase BBN-induced urinary bladder hyperplasia in either strain.

The results indicate that uracil acts as a weak co-carcinogen for renal pelvic in both strains of mice, but not for urinary bladder carcinogenesis.

Synergistic effects of phenacetin on renal pelvic and urinary bladder carcinogenesis by BBN

Since abuse of analgesics containing phenacetin has been strongly associated with development of urinary tract tumors [1–10], particularly of the renal pelvis, we investigated whether pretreatment with phenacetin influenced BBN carcinogenesis of the renal pelvis and urinary bladder of male NON/Shi mice.

Thirty-three male NON/Shi mice, 6 wk old, were given CA-1 diet and divided into four groups: groups 1, 2 and 4, had 10 mice each, and group 3 had 7 mice.

Mice of group 1 were given five treatments as follows: the first treatment was CA-1 diet containing 2% phenacetin (Phenactium, Japan Pharmacopoeia, from Maruishi Pharmacol Co. Ltd., Osaka, Japan) for 8 wk, the second treatment was basal diet containing no chemicals for 4 wk, the third treatment was CA-1 diet containing 2% phenacetin for 3 wk, the fourth treatment was drinking water containing 0.05% BBN for 4 wk, and the fifth treatment was water and diet containing no chemicals for 13 wk. Mice of group 2 were given the fourth treatment as group 1 and the other treatments without chemicals. Mice of group 3 were given the first, second, third and fifth treatments as group 1 and the fourth treatment without chemicals. Mice of group

TABLE IV

Synergistic effects of phenacetin on BBN-induced renal pelvic and urinary bladder carcinogenesis in male NON/Shi mice

Group	Phenacetin	BBN	Effective number of mice	Incidence of carcinomas	
				Renal pelvis[a] No.[c] (%)	Urinary bladder[b] No. (%)
1	+	+	10	5(50)	1(10)
2	−	+	10	0	0
3	+	−	7	0	0
4	−	−	10	0	0

[a]Renal pelvic carcinoma alone.
[b]Urinary bladder carcinoma alone.
[c]No. of mice with lesion.

4 were given all treatments without chemicals during experimental wk 32. The total observation period was 32 wk.

Macroscopically, 5 mice in group 1 had renal pelvic tumors and a mouse in group 1 had an urinary bladder tumor. No mice had urinary stones in the renal pelvis, ureter or urinary bladder.

Histopathological lesions of the renal pelvis and urinary bladder are summarized in Table IV. Group 1 had a 50% incidence of renal pelvic carcinomas, but groups 2–4 had no renal pelvic carcinomas. In group 1, 4 of the 5 renal pelvic carcinomas metastasized to several organs (incidences): lung (3 of 4 carcinomas), abdominal organs including liver and periaortic lymph nodes (2 of 4 carcinomas).

Group 1 had a 10% incidence of urinary bladder carcinomas, whereas groups 2–4 had no carcinomas. Moreover, an urinary bladder carcinoma in group 1 metastasized to liver and kidney.

The results indicate that pretreatment of phenacetin dramatically increased the carcinogenic activity of BBN in the renal pelvis rather than the urinary bladder of male NON/Shi mice.

Discussion

Of greatest interest in the present investigations was the observation that NON/Shi mice had significant susceptibility to renal pelvic carcinogenesis by BBN, and could serve as a suitable experimental model for the disease, DS/Shi and B6C3F1 mice were not susceptible. BBN-induced urinary bladder carcinogenesis has been shown to be influenced by genetic factors, such as strain and species [42–45], but renal pelvic carcinogenesis was not as clear. Before administration of BBN, ligation of one ureter in normal rats changed the target organ from the urinary bladder alone to the renal pelvis, ureter and urinary bladder [23]. Mechanical perforation of the renal pelvis also enhanced renal pelvic carcinogenesis by N-[4-(5-nitro-2-furyl)-2-thiazolyl]formamide in SD rats [18]. These reports suggested that urinary stagnation might play a role in renal pelvic carcinogenesis. Previously, we showed that SD-C rats, which have approximately a 90% incidence of spontaneous hydronephrosis, are sensitive to BBN-induced urinary tract carcinogenesis, especially of the renal pelvis [46]. NON/Shi mice also have spontaneous hydronephrosis and are susceptible to BBN-induced renal pelvic carcinogenesis, whereas DS/Shi and B6C3F1 mice are not. These results demonstrate a positive correlation between hydronephrosis and susceptibility to BBN-induced renal pelvic carcinogenesis.

The second interesting finding in the present work was that pretreatment with phenacetin synergistically enhanced BBN-induced renal pelvic carcinogenesis. Clinical and epidemiological studies have emphasized that phenacetin-containing analgesics produce human urinary tract carcinomas, especially of the renal pelvis [9–14]. Phenacetin has also shown promoting activity on renal pelvic and urinary bladder carcinogenesis of in rats [18, 47], with weak carcinogenic action toward urinary bladder carcinogenesis in rats and mice [16, 22]. A few investigators have suggested that phenacetin has clastogenic activity on rat peripheral lymphocytes [48] and mouse embryo cells [49] and micronucleus induction of mouse bone marrow polychromatic erythrocytes [50, 51].

The third interesting finding in the present work was that renal pelvic carcinomas in NON/Shi mice had higher incidences of metastasizing malignant carcinomas during an earlier period than urinary bladder carcinomas. A similar trend has been observed in patients with carcinomas of the urinary tract. Moreover, NON/shi mice have spontaneous relative hyperglycemia [37–39] and cell infiltration around the renal arteries and salivary gland ducts [52] which may be related to immunological abnormalities that might influence metastatic potential [53, 54]. Further investigations are needed to clarify the relationship between metastases of renal pelvic carcinomas and biological characteristics of NON/Shi mice.

The fourth interesting finding in the present work was that male-sex and old-age enhanced susceptibility to renal pelvic carcinogenesis by BBN in NON/Shi mice, whereas the phenomenon was not observed in urinary bladder carcinogenesis. The results suggest that aging male and female NON/Shi mice may increase the incidence of unilateral hydronephrosis, leading to urinary stagnation and carcinogen exposure. Clinical studies indicate that men have higher incidences of renal pelvic carcinomas than women [6–8]. Male mice have a higher susceptibility to BBN-induced urinary bladder carcinogenesis than female mice [55], and old male rats showed higher susceptibility to BBN-induced urinary bladder carcinogenesis than young and middle-aged male rats [56]. Moreover, old men are more sensitive to urinary bladder and renal pelvic carcinogenesis [5, 6, 57]. Thus, the factors of age and male-sex may play important roles in urinary tract carcinogenesis.

The fifth interesting finding in the present work was that uracil had weak co-carcinogenic potential on renal pelvic carcinogenesis in NON/Shi and ICR mice, but not on urinary bladder carcinogenesis. Male F344 rats showed susceptibility to the co-carcinogenic potential of uracil in renal pelvic and urinary bladder carcinogenesis by BBN [24, 25]. Uracil may form microcalculi in the renal pelvis of NON/Shi mice. The renal pelvic calculi of uracil enhanced activity of BBN to the epithelium in the renal pelvis.

Conclusion

The present studies report on the susceptibility on NON/Shi mice, having unilateral hydronephrosis, to renal pelvic and urinary bladder carcinogenesis by BBN compared with DS/Shi and B6C3F1 mice lacking spontaneous hydronephrosis. NON/Shi mice are sensitive to BBN-induced renal pelvic carcinogenesis, whereas the two other strains of mice are not. All strains of mice showed susceptibility to BBN urinary bladder carcinogenesis. Renal pelvic carcinogenesis in NON/Shi mice is strongly influenced by the factors of age and sex. Old male NON/shi mice had a 30% incidence of metastasizing renal pelvic carcinomas. Moreover, uracil showed weak co-carcinogenic activity for renal pelvic carcinogenesis in NON/Shi and ICR strains of male mice, but not for urinary bladder carcinogenesis. Phenacetin was dramatically synergistic for BBN-induced renal pelvic carcinogenesis in male NON-Shi mice, but for the bladder. Thus, NON/shi mice are a suitable model for renal pelvic carcinogenesis with high malignancy. It demonstrates a positive correlation between renal pelvic carcinogenesis and incomplete stagnation of urine containing the carcinogen.

Acknowledgements

We are indebted to Professor Dr. Nobuyuki Ito for helpful comments and critical review of this manuscript. We gratefully acknowledge the advice of Dr. Samuel M. Cohen in the preparation of this manuscript.

References

1 Ito N, Arai M, Sugimura S, Hirao K, Makiura S, Matayoshi K, Denda A. Experimental urinary bladder tumors induced by N-butyl-N-(4-hydroxybutyl)nitrosamine. Gann Monogr 1975;17:367–381.

2 Fukushima S, Masao H, Tsuda H, Shirai T, Hirao K, Arai M, Ito N. Histological classification of urinary bladder cancers in rats induced by N-butyl-N-(4-hydroxybutyl)nitrosamine. Gann 1976;67:81–90.

3 Ito N, Fukushima S. Promotion of urinary bladder carcinogenesis in experimental animals. Exp Pathol 1989;36:1–15.

4 Cohen SM, Friedell GH. Neoplasms of the urinary system. In: Foster HL, Small JD, Fox JG, eds. The mouse in biomedical research: Vol. IV. Experimental Biology and Oncology. New York: Academic Press, 1982; 440–444.

5 Riche JP, Shipley WU, Yagoda A. Cancer of the bladder. In: DeVita VT, Hellman S, Rosenberg SA, eds. Cancer: Principle & Practice of Oncology, 2nd Ed. Philadelphia: JB Lippincott, 1985;915–928.

6 Eagan JW Jr. Urothelial Neoplasms: Renal pelvis and ureter. In: Hill GS, ed. Uropathology, Vol. 2. New York: Churchill Livingstone, 1989;843–859.

7 Raghavan D, Russell P, Wong J, Pearson B, Malden LT. Urothelial malignancy of the upper tracts. In: Williama CJ, Krikorian JG, Green MR, Raghavan D, eds. Textbook of uncommon cancer. Oxford: Wiley, 1988;295–305.

8 Paulson DF, Perez CA, Anderson T. Cancer of the kidney and ureter. In: DeVita VT, Hellman S, Rosenberg SA, eds. Cancer: Principle & Practice of Oncology, 2nd Ed. Philadelphia: JB Lippincott, 1985;895–913.

9 Hultengren N, Lagergren C, Ljungquist A. Carcinoma of the renal pelvis in papillary necrosis. Acta Chir Scand 1965;130:314–320.

10 Angervall L, Bengtsson M, Zetterlund CG, Zsigmond M. Renal pelvic carcinoma in a Swedish district with abuse of phenacetin-containing drug. Brit J Urol 1969;41:401–405.

11 Johansson S, Angervall L, Bengtsson U, Wahlqvist L. Uroepithelial tumors of the renal pelvis associated with abuse of phenacetin-containing analgesics. Cancer 1974;33:743–753.

12 Propaczy P, Schramek P. Analgesic nephropathy and phenacetin-induced transitional cell carcinoma – analysis of 300 patients with long-term consumption of phenacetin-containing drugs. Eur Urol 1981;7:349–353.

13 McCredie M, Stewart JH, Ford JM, Taylor JS, Stewart JH. Analgesics and cancer of renal pelvis in New South Wales. Cancer 1983;55:220–225.

14 Palvio DHB, Andersen JC, Falk E. Transitional cell tumors of the renal pelvis and ureter associated with capillar sclerosis indicating analgesic abuse. Cancer 1987;59:972–976.

15 Johansson SL, Angervall L. Urothelial changes of the renal papillae in Sprague–Dawley rats induced by long term feeding of phenacetin. Acta Path Microbiol Scand Sect A 1976;84:375–383.

16 Isaka H, Yoshii H, Otsuji A, Koike M, Nagai Y, Koura M, Sugiyusu K, Kanabayashi T. Tumors of the Sprague-Dawley rat induced by long-term feeding of phenacetin. Gann 1979;70:29–36.

17 Johansson SL. Carcinogenicity of analgesics: the effect of long-term treatment of Sprague–Dawley rats with phenacetin, phenazone, caffeine and paracetamol. Int J Cancer 1981;27:521–529.

18 Anderstrom C, Johansson SL, Schultz L. The influence of phenacetin or mechanical perforation on the development of renal pelvic and urinary bladder tumors in FANFT-induced urinary tract carcinogenesis. Acta Path Microbiol Imunol Scand Sect A 1983;91:373–380.

19 Anderstrom C, Johansson SL. The combined effect of mechanical trauma and phenacetin or sodium saccharin on the rat urinary bladder. Acta Path Microbiol Immunol Scand Sect A 1983;91:381–389.

20 Kunze EK, Woltjen HH, Alberrecht H. Absence of a complete carcinogenic effect of phenacetin on the quiescent and proliferating urothelium stimulated by partial cystectomy. Urol Int 1983;38:95–103.

21 Johansson SL, Radio SJ, Saidi J, Sakata T. The effect of acetaminophen, antipyrine and phenacetin on rat urothelial proliferation. Carcinogenesis 1989;10:105–111.

22 Nakanishi K, Kurata Y, Oshima M, Fukushima S, Ito N. Carcinogenicity of phenacetin: long-term feeding study in B6C3F1 mice. Int J Cancer 1982;29:439–444.

23 Ito N, Makiura S, Yokota Y, Kamamoto Y, Hiasa Y, Sugihara S. Effect of unilateral ureter ligation on development of tumors in the urinary system of rats treated with N-butyl-N-(4-hydroxybutyl)nitrosamine. Gann 1971;62:359–365.
24 Shirai T, Tagawa Y, Fukushima S, Imaida K, Ito N. Strong promoting activity of reversible uracil-induced urolithiasis on urinary bladder carcinogenesis in rats initiated with N-butyl-N-(4-hydroxybutyl)nitrosamine. Cancer Res 1987;47:6726–6730.
25 Okumura M, Shirai T, Tamano S, Ito N, Yamada S, Fukushima S. Uracil-Induced calculi and carcinogenesis in the urinary bladder of rats treated simultaneously with N-butyl-N-(4-hydroxybutyl)nitrosamine. Carcinogenesis 1991;12:35–41.
26 Masui T, Mann AM, Garland EM, Cohen SM. Strong promoting activity by uracil on urinary bladder carcinogenesis and a possible inhibitory effect on thyroid tumorigenesis in rats initiated with N-methyl-N-nitrosourea. Carcinogenesis 1989;10:1417–1474.
27 Lalich JJ. Experimentally induced uracil urolithiasis in rats. J. Urol 1966;95:83–85.
28 Shirai T, Ikawa E, Fukushima S, Masui T, Ito N. Uracil-induced urolithiasis and the development of reversible papillomatosis in the urinary bladder of F344 rats. Cancer Res 1986;46:2062–2067.
29 Shirai T, Fukushima S, Tagawa Y, Okumura M, Ito N. Cell proliferation induced by uracil-calculi and subsequent development of reversible papillomatosis in the rat urinary bladder. Cancer Res 49;1989:378–383.
30 Yamamoto A, Shirai T, Kagawa M, Okamura T, Inoue K, Fukushima S. Mechanisms underlying uracil-induced bladder carcinoma development. Proc Jpn Cancer Assoc 1989;48:65 (in Japanese).
31 Mori S, Takeuchi Y, Toyama M, Makino S, Ohhara T, Hirota R, Nakai S. Amitrole-induced liver lesion in male NOD mice. Soc Jpn Toxicol Pathol 1987;3:57 (in Japanese).
32 Makino S, Kunimoto K, Muraoka Y, Mizushima Y, Katagiri K, Tochino Y. Breeding of a non-obese, diabetic strain of mice. Exp Anim 1980;29:1–13.
33 Tarui S, Tochino Y, Nonaka K (eds). Insulitis and Type I Diabetes: Lessons from the NOD mouse. Tokyo: Academic Press, 1986.
34 Hattori M, Buse JB, Jackson RA, Glimcher L, Dorf ME, Minami M, Makino S, Moriwaki K, Kuzaya H, Imura H, Strauss WM, Seidman JG, Eisenbarth GS. The NOD mouse: recessive diabetogenic genes in the major histocompatibility complex. Science 1986;231:733–735.
35 Nishimoto H, Kikutani H, Yamamura K, Kishimoto T. Prevention of autoimmune insulitis by expression of I–E molecules in NOD mice. Nature (London) 1987;328:432–434.
36 Lund T, Simpson E, Cooke A. Restriction fragment length polymorphisms in the major histocompatibility complex of the non-obese diabetic mouse. J Autoimmun 1990;3:289–298.
37 Ohgaku S, Morioka H, Yano S, Yamamoto H, Okamoto H, Tochino Y. NON mice: a new animal model for non-obese NIDDM. Diabetes Res Clin Pract 1985;Suppl 1:S415.
38 Kano Y. Insulin secretion in NOD (non-obese diabetic) and NON (non-obese non-diabetic) mice. J Kyoto Pref Univ Med 1988;97:295–308.
39 Sawa T, Ohgaku S, Morioka H, Yano S. Molecular cloning and DNA sequence analysis of preproinsulin genes in the NON mouse, an animal model of human non-obese, non-insulin-dependent diabetes. J Mol Endocrinol 1990;5:61–67.
40 Ohtani M, Tadano K, Nishio Y, Sato S, Sugimura T, Fukushima S, Niijima T. Sequential changes of mouse bladder epithelium during induction of invasive carcinomas by N-butyl-N-(4-hydroxybutyl)-nitrosamine. Cancer Res 1986;46:2001–2004.
41 Akaza H, Koiso K, Niijima T. Clinical evaluation of urothelial tumors of the renal pelvis and ureter based on a new classification system. Cancer 1987;59:1369–1375.
42 Hirose M, Fukushima S, Hananouchi M, Shirai T, Ogiso T, Takashi M, Ito N. Different susceptibilities of the urinary bladder epithelium of animal species to three nitroso compounds. Gann 1976;67:175–189.

43 Ohtani M, Kakizoe T, Sato S, Sugimura T, Fukushima S. Strain differences in mice with invasive bladder carcinomas induced by N-butyl-N-(4-hydroxybutyl)nitrosamine. J Cancer Res Clin Oncol 1986;112:107–110.
44 Nakanowatari J, Fukushima S, Imaida K, Ito N, Nagase S. Strain differences in N-butyl-N-(4-hydroxybutyl)nitrosamine bladder carcinogenesis in rats. Jpn J Cancer Res 1988;79:453–459.
45 Mori S, Murai T, Takeuchi Y, Toyama M, Makino S, Konishi T, Hayashi Y, Kurata Y, Fukushima S. Dose response of N-butyl-N-(4-hydroxybutyl)nitrosamine on urinary bladder carcinogenesis in mutant ODS rats lacking L-ascorbic acid synthesizing ability. Cancer Lett 1990;49:139–145.
46 Mori S, Takeuchi Y, Makino S, Kurata Y, Fukushima S. Induction of renal pelvic and ureteral tumors in SD-C rats by N-butyl-N-(4-hydroxybutyl)nitrosamine. Proc Jpn Cancer Assoc 1987;46:58 (in Japanese).
47 Nakanishi K, Fukushima S, Shibata M-A, Shirai T, Ogiso T, Ito N. Effect of phenactin and caffeine on the urinary bladder of rats treated with N-butyl-N-(4-hydroxybutyl)-nitrosamine. Gann 1978;69:395–400.
48 Granberg-Ohman I, Johansson SL, Hjerpe A. Sister-chromatid exchanges and chromosomal aberrations in rats treated with phenacetin, phenazone and caffeine. Mutation Res 1980;79:13–18.
49 Patierno SR, Lehmann NL, Henderson BE, Landolph JR. Study of the ability of phenacetin, acetaminophen, and aspirin to induce cytotoxicity, mutation, and morphological transformation in C3H/10T1/2 clone 8 mouse embryo cells. Cancer Res 1989;49:1038–1044.
50 Sutou S, Kondo M, Mitsui Y. Effects of multiple dosing of phenacetin in the micronucleus test. Mutation Res 1990;234:183–186.
51 Sutou S, Mitui Y, Toda S, Seilima M, Kawasaki K, Ando N, Kawata T, Abe S-I, Iwai M, Arimura H. Effect of multiple dosing of phenacetin on micronucleus induction: a supplement to the international and Japanese cooperative studies. Mutation Res 1990;245:11–14.
52 Hadano H, Suzuki S, Tanigawa K, Ago A. Cell infiltration in various organs and dilation of the urinary tubule in NON mice. Exp Anim 1988;37:479–483 (in Japanese).
53 Brunson KW, Goldfarb RH. Immunosuppression by metastatic tumors. In: Herbermann RB, ed. Cancer growth and progression: Influence of the host on tumor development. Dordrecht/Boston, Kluwer, 1989;133–138.
54 Feldman M, Eisenbach L. What makes a tumor cell metastatic? Sci Am 1988; November: 40–47.
55 Bertram JS, Craig AW. Specific induction of bladder cancer in mice by butyl-(4-hydroxybutyl)nitrosamine and the effects of hormonal modification on the sex differences in response. Europ J Cancer 1972;8:587–594.
56 Fukushima S, Shibata M-A, Tamano S, Ito N, Suzuki E, Okada M. Aging and urinary bladder carcinogenesis induced in rats by N-butyl-N-(4-hydroxybutyl)nitrosamine. J Natl Cancer Inst 1987;79:263–267.
57 Ito N. Urinary bladder cancer: growth, progression and the modifier. Tr Soc Pathol Jpn 1986;75:3–37 (in Japanese).

Subject Index